A Simplex Approach to Learning, Cognition, and Spatial Navigation:

Emerging Research and Opportunities

Pio Alfredo Di Tore
University of Salerno, Italy

A volume in the Advances in Educational Technologies and Instructional Design (AETID) Book Series

IGI Global
DISSEMINATOR OF KNOWLEDGE

Published in the United States of America by
IGI Global
Information Science Reference (an imprint of IGI Global)
701 E. Chocolate Avenue
Hershey PA, USA 17033
Tel: 717-533-8845
Fax: 717-533-8661
E-mail: cust@igi-global.com
Web site: http://www.igi-global.com

Library of Congress Cataloging-in-Publication Data

Names: Di Tore, Pio Alfredo, 1973- author.
Title: A simplex approach to learning, cognition, and spatial navigation : emerging research and opportunities / by Pio Alfredo Di Tore.
Description: Hershey, PA : Information Science Reference, 2018.
Identifiers: LCCN 2017002934| ISBN 9781522524557 (hardcover) | ISBN 9781522524564 (ebook)
Subjects: LCSH: Space perception in children. | Spatial behavior. | Cognitive learning. | Child psychology.
Classification: LCC BF723.S63 D52 2018 | DDC 155.4/13752--dc23 LC record available at https://lccn.loc.gov/2017002934

This book is published in the IGI Global book series Advances in Educational Technologies and Instructional Design (AETID) (ISSN: 2326-8905; eISSN: 2326-8913)

British Cataloguing in Publication Data
A Cataloguing in Publication record for this book is available from the British Library.

Advances in Educational Technologies and Instructional Design (AETID) Book Series

ISSN:2326-8905
EISSN:2326-8913

Editor-in-Chief: Lawrence A. Tomei, Robert Morris University, USA

MISSION

Education has undergone, and continues to undergo, immense changes in the way it is enacted and distributed to both child and adult learners. From distance education, Massive-Open-Online-Courses (MOOCs), and electronic tablets in the classroom, technology is now an integral part of the educational experience and is also affecting the way educators communicate information to students.

The **Advances in Educational Technologies & Instructional Design (AETID) Book Series** is a resource where researchers, students, administrators, and educators alike can find the most updated research and theories regarding technology's integration within education and its effect on teaching as a practice.

COVERAGE

- Online Media in Classrooms
- Collaboration Tools
- Digital Divide in Education
- Game-Based Learning
- Web 2.0 and Education
- Higher Education Technologies
- K-12 Educational Technologies
- Instructional Design Models
- Classroom Response Systems
- Curriculum Development

The Advances in Educational Technologies and Instructional Design (AETID) Book Series (ISSN 2326-8905) is published by IGI Global, 701 E. Chocolate Avenue, Hershey, PA 17033-1240, USA, www.igi-global.com. This series is composed of titles available for purchase individually; each title is edited to be contextually exclusive from any other title within the series. For pricing and ordering information please visit http://www.igi-global.com/book-series/advances-educational-technologies-instructional-design/73678. Postmaster: Send all address changes to above address.

Titles in this Series

For a list of additional titles in this series, please visit:
https://www.igi-global.com/book-series/advances-educational-technologies-instructional-design/73678

Optimizing STEM Education With Advanced ICTs and Simulations
Ilya Levin (Tel Aviv University, Israel) and Dina Tsybulsky (Tel Aviv University, Israel)
Information Science Reference • ©2017 • 344pp • H/C (ISBN: 9781522525288) • US $180.00

Handbook of Research on Classroom Diversity and Inclusive Education Practice
Christina M. Curran (University of Northern Iowa, USA) and Amy J. Petersen (University of Northern Iowa, USA)
Information Science Reference • ©2017 • 552pp • H/C (ISBN: 9781522525202) • US $235.00

Handbook of Research on Instructional Systems and Educational Technology
Terry Kidd (University of Houston-Downtown, USA) and Lonnie R. Morris, Jr. (The Chicago School of Professional Psychology, USA)
Information Science Reference • ©2017 • 493pp • H/C (ISBN: 9781522523994) • US $265.00

Applying the Flipped Classroom Model to English Language Arts Education
Carl A. Young (North Carolina State University, USA) and Clarice M. Moran (Kennesaw State University, USA)
Information Science Reference • ©2017 • 277pp • H/C (ISBN: 9781522522423) • US $185.00

Digital Tools and Solutions for Inquiry-Based STEM Learning
Ilya Levin (Tel Aviv University, Israel) and Dina Tsybulsky (Tel Aviv University, Israel)
Information Science Reference • ©2017 • 371pp • H/C (ISBN: 9781522525257) • US $180.00

Exploration of Textual Interactions in CALL Learning Communities...
Jonathan R. White (Högskolan Dalarna, Sweden)
Information Science Reference • ©2017 • 195pp • H/C (ISBN: 9781522521426) • US $120.00

Mobile Technologies and Augmented Reality in Open Education
Gulsun Kurubacak (Anadolu University, Turkey) and Hakan Altinpulluk (Anadolu University, Turkey)
Information Science Reference • ©2017 • 366pp • H/C (ISBN: 9781522521105) • US $185.00

For an enitre list of titles in this series, please visit:
https://www.igi-global.com/book-series/advances-educational-technologies-instructional-design/73678

IGI Global
DISSEMINATOR OF KNOWLEDGE

701 East Chocolate Avenue, Hershey, PA 17033, USA
Tel: 717-533-8845 x100 • Fax: 717-533-8661
E-Mail: cust@igi-global.com • www.igi-global.com

Table of Contents

Foreword

HUMAN MOVEMENT: THE SPATIAL LANGUAGE OF THE BRAIN

This work aims to provide a "point of view" on the relationship between space, human movement and cognition, starting from a hypothesis developed by Alain Berthoz:

My hypothesis...is that the mental tools elaborated during evolution to solve the many problems posed by the need to advance into space have also been used for the higher cognitive functions: memory and reasoning, relationship with others and also creativity. (Berthoz, 2011b)

Space, in short, is not just a matter of geometers, as the movement is not just about coaches, doctors and physiotherapists.

According to the latest neuroscience perspective, mechanisms specialized in action-oriented visuospatial perception and processing involve cognitive mechanisms, or, if preferred, "the brain uses a spatial language" (Berthoz, 2011a).

Space, in this view, cannot be separated, according to the work of Merleau-Ponty (1962), from the perceiver: You can know reality only in experiential learning.

Perception, here, is not a moment separate from action in an algorithmic process (perception-elaboration-action) but is part of action, indeed, it is a function of action. There is no perception, there is no vision of the world that does not refer to the moving body in space.

More precisely, the body in action resolves complexity in a process of perception-action which, according to the most recent findings of neuroscience, is reversed with respect to the cognitivist paradigm. Leman (2008) summarizes:

What happens in perception can be understood in terms of action. (Berthoz, 1997; Decety & Jackson, 2004; Jeannerod, 1994; Prinz & Hommel, 2002)

Berthoz writes about the:

Reversal of the classical description of the mechanisms of perception and action places the intentional and goal-oriented subject at the origin of the process. The subject builds his world according to his basic needs and action tools. This view has also been promoted by Bergson and Husserl. (Berthoz, 2008a)

The ability to have a space view is therefore not only the necessary condition to move from point A to point B, but it is the foundation of the ability to consider the world in different ways, to change not only the point of view but also the interpretation of reality, the attribution of values, the tolerance of difference, making a decision.

These premises require the need for an approach that is not simply multi disciplinary but, upstream, trans-disciplinary. A crisscrossed landscape, to name.

If we trust the peer review reviews, the work that follows succeeds in this business, its strengths are: innovation; far-reaching content; current research; and the fact that it represents a new perspective across disciplines.

It spans decades, research methodology, and conceptual frameworks. It has practical significance for the fields of psychology, neurology, philosophy, and gaming. The work is innovative, bridging fields and concepts across disciplines. There would be multiple audiences for this work: schools of art; philosophy; psychology; and neurology all being interested.

However, the span of fields (gender, videogames, child perspective taking, Sport sciences, special needs students, empathy) requires a clear link, and the author tries to supply it in the introduction (where successfully aligns the fields to a specific context), but also throughout the individual chapters. Each of these fields comprises very specific content and vocabulary.

In summary, the theoretical framework in this work is at the basis of a path of innovative research that is in an ideal crossroads between several disciplines (philosophy, education, sport sciences, neuroscience), recognizing a priori that the deviations that are on the path are all deserving of attention.

The road, impervious, chosen by the project is that of experimental verification, the concrete attempt to equip itself with appropriate tools for the multi-faceted nature of the field of investigation, adapting them from other fields or, as in the present case, developing them ex-novo.

Gaetano Raiola

REFERENCES

Berthoz, A. (1997). *Le sens du mouvement* (Vol. 223). Odile Jacob Paris.

Berthoz, A. (2008a). *Neurobiology of "Umwelt": How Living Beings Perceive the World*. Springer.

Berthoz, A. (2011a). Fondements cognitifs de la perception de l'espace. In J.-F. Augoyard (Ed.), Faire une ambiance = creating an atmosphere: actes du colloque international Grenoble 10-12 septembre 2008 (pp. 121-132). Grenoble: A la croisée.

Berthoz, A. (2011b). *La semplessità*. Torino: Codice.

Decety, J., & Jackson, P. L. (2004). The functional architecture of human empathy. *Behavioral and Cognitive Neuroscience Reviews*, *3*(2), 71–100. doi:10.1177/1534582304267187 PMID:15537986

Jeannerod, M. (1994). The representing brain: Neural correlates of motor intention and imagery. *Behavioral and Brain Sciences*, *17*(02), 187–202. doi:10.1017/S0140525X00034026

Leman, M. (2008). *Embodied Music: Cognition and Mediation Technology*. MIT Press.

Merleau-Ponty, M. (1962). *Phenomenology of Perception: An Introduction*. Routledge & Kegan Paul.

Prinz, W., & Hommel, B. (2002). *Common mechanisms in perception and action: Attention and performance XIX*. Oxford University Press.

Preface

I KNOW WHAT YOU SEE

Space, perspective-taking, empathy, learning, video games, corporeality, sports, orientation are the key words of this work. Aim of this introduction is to outline a conceptual framework in which these apparently heterogeneous elements find an organic placement.

The project presents a research perspective taking, aimed to investigate two specific issues: the age when the child leaves, as a rule, perceptive egocentrism, and gender differences in the management of spatial reference systems.

The work is part of a busy line of investigation that, from 1948 - the year of publication of "la représentation de l'espace chez l'enfant" - to the recent Nobel prize for medicine awarded in the field of cognitive neuroscience of space, has saw the contributions of scholars from different traditions and different fields of study (Piaget, Berthoz, Baron-Cohen, Rochat, Frith, DeVignemont, Hughes, Fink, Decety, Vogeley, Burgess, O'Keefe, just to name a few).

One of the major critical points discussed later, identified by scholars, is the choice of survey tools. In research involving children and teenagers, involvement, motivation, and accurate understanding of the nature of the task to be performed by individuals are fundamental variables. Piaget's research, for example, was later criticized because of the lack of interest that the experiment arouses in children.

The contribution (hopefully) innovative of this work is that it has addressed the issue with a non-abused instrument for this field: the videogame.

The videogame, in fact, offers what J.P. Gee has defined an empathetic embodiment (Gee, 2004a), an immersive process in a narrative, interactive and perceptual context, and, in the case of the research in question, greatly simplifies the understanding of the task (*what should I do?*) and the motivation to the task (*why should I do that?*). Learning through video games is, for Gee, significant and meaningful because it is based not on definition, but on direct experience and simulation within specific semantic domains. According to Gee, good video games incorporate good learning principles (Gee, 2004b). Challenge and learning are a large part of what makes good video games motivating and entertaining. The first of these principles, simply, is "What you learn when you learn to play a video game is just how to play the game", which responds exactly, as we shall see, to a large part of the criticisms made to Piaget's experiments.

The choice of instrument appeared, actually, natural, on the one hand because the current generation of tools for the design and development of videogames offers tools for the management of space and three-dimensional environments that simulate the physical laws, and secondly because research on the point of view in the game, in the cinema, in the narrative, in the fairy tale (all of which are elective affinities with the videogame) looks like solid ground, even if vast and complex that could provide valuable references.

The assumption that virtual context is suitable for studies of this nature was experimentally established, among the others, by Alain Berthoz:

We designed a behavioral study in which participants interacted spontaneously with a life-sized virtual tightrope walker walking forward, backward and leaning to her left or right on a rope. Here, we report results showing that participants automatically embodied the avatar's leaning movements. (Berthoz & Thirioux, 2010)

Chapters 1-4 describe the conceptual framework underlying the research and offer a review of the scientific literature.

The design and development of the video game development, research design, and the results obtained, are illustrated in Chapter 5, "IkeWYSe: I Know What You See."

Chapter 6, "Vicariance, Complexity, and Spatial Navigation," is developed by introducing the concept of Vicariance, quickly describing the orienteering as a sport, and, in the light of the review of the scientific literature on the cognitive processes involved in spatial navigation and in System Management of spatial reference, concludes with a proposal for the spread of orienteering as an effective teaching practice in a training environment oriented to the inclusion of pupils presenting Special educational Needs (Corona, 2014).

CONCEPTUAL FRAMEWORK

Rehabilitation of Descartes

The Cartesian thought, which traditionally has never gone hand in hand with neuroscience (Damasio, 2000), reveals formidable similarities with these about the concept of space - or, rather, presents insights and theories that find confirmation in the latest research in neuroscience - when Descartes states that space is an extension of the body, i.e., the body itself, so it is wrong to talk of space and bodies, and it is more correct to speak of space of bodies.

The idea that we have of the body is included in the idea we have of space. The merit of the rationalists and, in particular, of the empiricists, is, ultimately, to have placed the space in the cognitive world of the subject, giving it an epistemological perspective. The point of view of the empiricists is the first to frame the space from the perspective of the acting subject, the perspective that underlies the concept of space - of that part of the space that is part of the perceptual bubble (Umwelt) of the subject – as an event of the perception-action process, such as psychomotor functional event (Llinás, 2009).

The perception-action process is understood here according to the new paradigm developed in neuroscience, which appears reversed with respect to "traditional" implant. Perception and action are not currently described as distinct, sequential and discrete phases, but as an ongoing process in which perception is function of action, it is not subordinated to a contemplative vision of a (presumed) objective reality, but it is structured for the action, motivates and prepares it. The generative core of this process is the body in action, with intent and aimed at a goal (goal-oriented). In other words:

the subject builds his world according to his basic needs and action tools. This view has Also been promoted by Bergson and Husserl. (Berthoz, 2008)

A quick historical review on the idea of space is present in Chapter 1, "Space, Perception, Action," in which neuroscientific perspective of perception-action is presented (perspective that today provides empirical evidence, but whose premises had been built, with considerable advance, from the phenomenological tradition, with particular reference to the work of Merleau-Ponty), and introduces the concept of Umwelt, in the sense developed at the Collège de France.

Space Representation

Alain Berthoz in a flash introduces as follows the inseparable link between space and the cognitive processes:

Space is not just a matter of geometers and architects. It has long been used for memory. The monks-have always used mental spaces to memorize the concepts, data, ideas and priests in the parishes used mental spaces to make the paths and also find several sermons on Sunday. Space is not simply a matter of memory, it is a technique used to find new solutions in cognitive functions. (Belthoz, 2011a)[1]

The brain uses a spatial language, the mechanisms specialized in perception and action-oriented visuospatial processing involve cognitive mechanisms.

In the hypothesis developed at the Collège de France, some cognitive mechanisms would be the result of spatial processing. The hypothesis is that:

Mental tools developed during evolution to solve the many problems posed by the advancement in space have also been used for higher cognitive functions: memory and reasoning, the relationship with each other and also the creativity. (Berthoz, 2011b)

The link between space, cognition, learning is structured, non-mechanical, sometimes subtle, as we can verify encountering discrepancy between lived space and perceived space in many neurological or psychiatric disorders.

Berthoz, in this regard, is explicit, and suggests:

A new hypothesis: the manipulation of spatial references would be a transnosographic tract in various neurological and psychiatric disorders. I mean that each of the following diseases - autism, schizophrenia, epilepsy, spatial anxiety, Parkinson's disease - affects structures involved in the manipulation of different spatial referents. Consequently we will find in these diseases the deficiency of specific symptoms in each: some will produce more perceptual and motor problems, other orientation problems, others a difficulty in changing the point of view. These deficits in vicariance generate stereotyped behaviors or, as we shall see, the difficulty of interacting with others. (Berthoz, 2014)

The ability to have a global vision of space (a coherent functional representation) is associated with the ability to view the world in different ways, to change not only the views but also the interpretation of reality, the assignment of values, the ability to tolerate difference, to make a decision.

Jean Piaget claimed, with the theory of stages of development of spatial thinking, that the child acquires the ability to manipulate the points of view of space around seven / eight years of age. Only at this age, for Piaget:

The intellectual space will be built, capable of definitively taking it over the perceptible space and allowing a manipulation of points of view which is not merely spatial but which in fact also associates a whole series of mechanisms of representation, And even semantics. (Piaget & Inhelder, 1948)[2]

Commenting Piaget, and in particular the task of the three mountains, the most famous experimental Piaget's test on the representation of space in the child, Trisciuzzi Zappaterra summarize:

A child, however, still immersed in a egocentric point of view, find difficult to choose a figure out of its narrow point of view. Going out from self, therefore, means being able to see the reality (not only spatial, but also a mental reality, linguistic or otherwise) from several points of view, however other than their own. (Trisciuzzi & Zappaterra, 2011)

To get out perceptual egocentrism is not, however, reducible to the mechanical assumption of another's position in space. Building a coherent representation of space implies the ability to make a:

Mental rotation on ourself, in relation to the environment or to an object of the environment, maintaining the main perspective of the environment in question. (Berthoz, 2011b)

It implies, in other words, the ability to take another's perspective (perspective taking) without abandoning completely his own, a form of out-of-body experience that separates us from our bodies and let us "surf" the bodies of others through our "second self" or "mental double" (Berthoz & Petit, 2006), or "doppelgänger" (Brugger, 2002).

In practice, it comes to be at the same time oneself and the other. Just this is the basic feature, for Alain Berthoz, empathy.

Spatial reference systems competing in the representation of space, Piaget's conceptual framework, scientific debate on the "three mountains" and spatial theory of empathy are the topic of Chapter 2, "Space Representation in Children: From Piaget to Neuroscience."

Space Invaders

As mentioned, the tool designed and developed for data collection, in this research, is a video game. The representation of space and the management's point of view in video games become, then, very important elements to the paper.

Many video games take the first-person perspective as the dominant perspective (FPC - First Person Camera) but at the same time allow you to change the perspective necessary (TPC - Thirdh Person Camera).

The habit to switch between the different perspectives is established and deemed effective among gamers. This consideration, easily verified by interviewing players or browsing dedicated magazine, is particularly relevant to this research. The gamer effectively masters different perspectives, is able to quickly select the most advantageous perspective, normally uses systems of representation (maps), builds functional representations of virtual spaces on the basis of the experiences on the screen.

It can be concluded that many contemporary computer games allow their players an ever greater amount of control over the spatial perspective(s) used in the presentation of the game space.... Finally, it may be noted that while players generally like the opportunity to take control of the camera, they rarely use the possibility to change the default point of view. This has to do with the fact that the default point of view is often best suited to the interaction with the game space required by the game. (Thon, 2009)

The use of points of view in video games and the similarities / differences between film camera and interactive ones are the subject of Chapter 4, "Space and Media: Digital Umwelten."

REFERENCES

Berthoz, A. (2008). *Neurobiology of "Umwelt": How Living Beings Perceive the World.* Springer.

Berthoz, A. (2011a). Fondements cognitifs de la perception de l'espace. In J.-F. Augoyard (Ed.), Faire une ambiance = creating an atmosphere: actes du colloque international Grenoble 10-12 septembre 2008 (pp. 121-132). Grenoble: A la croisée.

Berthoz, A. (2011b). *La semplessità.* Torino: Codice.

Berthoz, A. (2014). *La vicarianza. Il nostro cervello creatore di mondi.* Torino: Codice.

Berthoz, A., & Petit, J.-L. (2006). *Phénoménologie et physiologie de l'action.* Odile Jacob.

Berthoz, A., & Petit, J. L. (2006). *Physiologie de l'action et Phénoménologie.* Paris: Odile Jacob.

Berthoz, A., & Thirioux, B. (2010). A Spatial and Perspective Change Theory of the Difference Between Sympathy and Empathy. *Paragrana, 19*(1), 32–61. doi:10.1524/para.2010.0003

Brugger, P. (2002). Reflective mirrors: Perspective-taking in autoscopic phenomena. *Cognitive Neuropsychiatry, 7*(3), 179–194. doi:10.1080/13546800244000076 PMID:16571536

Corona, F. (2014). *Special Educational Needs.* Roma: Aracne Editrice.

Damasio, A. R. (2000). *L'errore di Cartesio.* Editmabi.com.

Gee, J. P. (2004a). *Videogames: Embodied empathy for complex systems.* Academic Press.

Gee, J. P. (2004b). *What Video Games Have to Teach Us About Learning and Literacy.* Palgrave MacMillan.

Llinás, R. R. (2009). Umwelt: A Psychomotor Functional Event. In A. Berthoz (Ed.), *Neurobiology of "Umwelt"* (pp. 29–37). Berlin: Springer. doi:10.1007/978-3-540-85897-3_4

Piaget, J., & Inhelder, B. (1948). *La représentation de l'espace chez l'enfant*. Paris: Presses Universitaires De France.

Thon, J.-N. (2009). Perspective in Contemporary Computer Games. *Point of View, Perspective, and Focalization: Modeling Mediation in Narrative*, 279-300.

Trisciuzzi, L., & Zappaterra, T. (2011). Dislessia, disgrafia e didattica inclusiva. *Annali Della Pubblica Istruzione*, *2*, 51–76.

ENDNOTES

1 L'espace, ce n'est pas seulement une affaire de géomètres et d'architectes. Il est utilisé depuis longtemps pour la mémoire. Depuis toujours les moines utilisent des espaces mentaux pour stocker des concepts, des données, des idées, et les curés dans les paroisses utilisaient ces espaces mentaux pour faire des cheminements et même trouver des serinons différents tolls les dimanche. L'espace n'est pas simplement un lieti de mémoire, c'est aussi une technique utilisée depuis toujours pour trouver des solutions nouvelles dans les fonctions cognitives. (Berthoz, 2011a)

2 l'espace intellectuel sera construit, capable de l'emporter définitivement sur l'espace perceptible et de permettre une manipulation des points de vue qui n'est pas simplement spatiale mais qui, en fait, associe aussi toute une série de mécanismes de représentation, et même sémantiques. (Piaget & Inhelder, 1948)

Chapter 1
Space, Perception, Action

ABSTRACT

This paper focuses on the concept of action and on the reversal of the process of perception-action operated by the phenomenological tradition (Merleau-Ponty), which has been confirmed in more or less recent neuroscientific evidences (Berthoz, Decety, Jeannerod). The text presents a quick historical overview on the concepts of space, perception, action, introducing the concept of Umwelt. The umwelt is, a dynamic, interactive concept that defines the relations between the physical world and living organisms, and constitutes the basis and the assumption of intersubjectivity. The subject builds up his world in accordance with his basic necessities and his tools of action. The chapter introduces the perspective of Alain Berthoz, who proposes a vision in which the subject navigates in his own umwelt led by a series of simplifying principles that optimize the process of perception - action and minimize the need for computation. This is coherent with the vision of Gell-Mann, according to whom an adaptive complex system (such as, for example, a living being) receives a data stream, and identifies the perceived regularity in the data stream, compresses their description in a schema, and then uses this schema for the description, the prediction or the action. According to this idea, the body in action resolves complexity in a process of perception-action which is reversed with respect to the cognitivist paradigm: "what happens in perception can be understood in terms of action".

DOI: 10.4018/978-1-5225-2455-7.ch001

THINKING SPACE

Questioning the nature of space, on the relationship between space and time (and on connected paradoxes) is in the DNA of Western culture since the time of Zeno.

What is space? What is time? Do they exist independently of the things and processes in them? Or is their existence parasitic on these things and processes? Are they like a canvas onto which an artist paints; they exist whether or not the artist paints on them? Or are they akin to parenthood; there is no parenthood until there are parents and children? That is, is there no space and time until there are things with spatial properties and processes with temporal durations? These questions have long been debated and continue to be debated. (Norton, 2004)

DiSalle, in *Understanding time-space*, indicates how the recent reflection on space constitutes a fil rouge, the continued debate from Newton to Einstein:

When Newton appeals to absolute space, he does not advance any theses about the ontology of space-time. Rather the postulation of absolute space and time is inspired by empirical reasoning about motion. This theme unites Newton with later physicists: At the very least, we can identify a common metaphysical principle uniting general relativity with special relativity and Newton's theory: space-time is an objective geometrical structure that expresses itself in the phenomena of motion. (DiSalle, 2006)

The philosophical debate on the ontology of space, in a nutshell, was hinged on two specular positions: an idea of absolute space, according to which space and time exist independently from the objects and object relations (or, more radically, according to which space and time exist), and an idea of relative space, according to which that the existence of space and time is linked to the objects and relationships between objects (or, more radically, according to which space and time do not exist) (DiSalle, 2006).

The Kantian position, that saw space and time as a priori frameworks, was the subject of further analysis of Poincare and Einstein, and grossed, as we shall see, the criticism of Piaget.

For Kant:

Space is a necessary a priori representation that underlies all outer intuitions. One can never forge a representation of the absence of space, though one can quite well think that no things are to be met within it. It must therefore be regarded as the condition of the possibility of appearances, and not as a determination dependent upon them, and it is an a priori representation that necessarily underlies outer appearances. (Kant, 1855)

Historically, only with the explosion of the Renaissance ferments the extraordinary mind of Giordano Bruno will propose an image of space as infinite being, uniform, indivisible, all-inclusive, not understood, opening the way to the conception of absolute space of Newton as sensorium Dei.

The ancient-medieval age, pervaded by subtle doctrinal disputes, we had intended, essentially, as a container of all material objects, without prejudice to its three-dimensional concept developed in the geometry of Thales, then codified by Euclid.

Eleatic thought, first, projected the epistemological problem in space. In fact, in Parmenides dualism, although the reality consists of the sensitive phenomena is illusory, lives of apparent life, the process of knowledge is aimed towards the only true reality, the being, which is not visible, it is not tangible, it can only be thinkable. Space, therefore, is tinged with contradiction, because each thing is what it is, because it is not each other. Therefore phenomenal existence is apparent, as is contradictory to the time and, therefore, the perception of all that changes in time. The space, therefore, as it will be for Plato, is an empty receptacle, invisible, distinct from the forms it takes, unchanging (unlike the forms), as opposed to being, being what it is not. It can only be guessed. Intuition, then, of course, depends on the constituent elements of the subject.

In order to arrive, then, to configure the nature of this intuition must wait for Kant, for which the space is a representation a priori that is based on internal intuitions. It is pure intuition. In this it extends and diversifies the fixity of the Aristotelian categories, as many are those who represent the sensible intuition of each individual, through which organizes, independently and subjectively, the phenomenon.

Fundamental is the assumption that space is *the condition of possibility of phenomena* (Kant, 1855).

It was, however, the explosion of rationalism (Descartes, Leibnitz) and the consequent empiricism (Locke, Berkeley), to stimulate, combating subjectivism induced, Kant's reflection on the intended space as transcendental condition of knowledge and to return it to the metaphysical sphere.

During the nineteenth and twentieth century Kant's view, both from the philosophical point of view, and especially on the objective basis of new scientific discoveries, was literally demolished.

The condition of possibility of phenomena is present, however, in some sense, in existentialist thought, especially Heidegger.

Identifying the space with the world, it elevates the spatiality to a set of possibilities. The spatiality is not only in things, but also in humans (being-there). Neither space is subject, nor the world is in space. Space is in the world because in it is the subject. The space is a condition of being-there, is presented as a place of call to action.

Recent discoveries in neuroscience have finally established the indissolubility of the triad space-mind-body.

On closer reflection, essentially, beyond the positions, so to speak, "ideological", little has changed since the Parmenidean reflection, focused on the obscure conception of being.

The phenomenological field of possibilities is the center from which to connote the spatiality.

Advanced research in neuroscience also, moreover, reflects on empirical data, extending the inexhaustible investigation field on being as a substance.

PERCEPTION, ACTION

The neuroscientific investigation does not take the space as a category, but frames it from the point of view of brain:

In treating space, the problem for the brain is the multiplicity of spaces. There is no single "space", there is a great variety of spaces. Secondly, perceive the space is not equivalent to perceive the geometry, but to perceive the movement. (Berthoz, 2011a)[1]

The brain, in this perspective, it is very far from the idea of processor of stimuli coming from the senses, but is a creator of worlds, a reality emulator.

Our reality emulator acts primarily as the prerequisite for coordinated, directed motricity; it does so by generating a predictive image of an event to come that causes the creature to react or behave accordingly. (Llinás, 2002)

A position, this, that Llinás does not hesitate to describe as "a brain-centric perspective", in which the brain is not:

An open system…that accepts inputs from the environment, processes them, and returns them to the world reflexively regardless of their complexity", ma è un "closed system modulated by the senses", cioè un "self-activating system, whose organization is geared toward the generation of intrinsic images, capable of emulating reality (generating emulative representations or images) even in the absence of input from such reality, as occurs in dream states or daydreaming. From this one may draw a very important conclusion. This intrinsic order of function represents the fundamental, core activity of the brain. This core activity may be modified (to a point!) through sensory experience and through the effects of motor activity." (Llinás, 2002)

The brain, in short, does not compute anything, not processing information, at least not in the sense of algorithmic data manipulation that is the main feature of Alan Turing universal (Millican & Clark, 1996; Turing, 1947).

Leman (2008) summarizes:

What happens in perception can be understood in terms of action. (Berthoz, 1997; Decety & Jackson, 2004; Jeannerod, 1994; Prinz & Hommel, 2002)

Berthoz is more explicit:

We base on the action, and not on representation, our conception of the organism. The perception does not represent the world as it is, but the structure in the Umwelt…. There is no perception of the world that does not refer in any way to the body which acts. (Berthoz, 2008b)

In the words of Berthoz you can find a definitive overcoming of every cognitivist hypothesis:

The brain is not a computer: a functionalistic concept of thinking as a symbolic computation is revealed (rather than developing superficially), which reveals a purely logical analysis unconnected with biological realizations. (Berthoz & Petit, 2006)[2]

In the case of perspective taking, our "emulator" worlds performs a mental déplacement, a simulated rotation of the body in space, a mental manipulation on the body and objects of the environment through a process that Berthoz defines of "mise à jour spatiale."

Berthoz cautions on the use of the term manipulation:

Indeed, we constantly change references. I use the term "manipulation of spatial references" to describe this action, although it is a mental and non-manual operation. In my opinion, such a richness of the repertoire of potential references allows functional vicariance (Berthoz, 2013)[3]

Vicariance is, in this sense, the Ariadne's thread that allows us to move in the space without end that stands between perception and action, conceptual territory and place of research that has been revealed in all its vastness only by overcoming any neopositivistic remaining and only with the collapse of the discretization (perception - processing - action as successive, separated moments) made by the computational metaphor.

Precisely in this arc infinite between perception and action, in the moment of the decision as a constant search for a perpetually renegotiated stability between reduction of complexity and production of meaning, the brain human produces a continue boost to change point of view, laying the foundations of intersubjectivity.

UMWELT

Umwelt is the term used by von Uexküll (von Uexküll, 1934) to indicate the "perceptual world".

Animals and humans are...prisoners of a specific repertoire of possible perceptions of their Umwelt. This term...indicates that the animal makes use of their environment, and not vice versa. In fact Uexküll believes that living organisms are constituted by a set of components that originate everything by combining, but which are structured from an aim, a real intention that determines the parts in a centrifugal manner, and not centripetal. It is the action that gives meaning to the artifacts in the sense of "art products": objects, tools, man-made machines.... The meaning is always given by the act that we are going to accomplish with the object, which simplifies its perception and especially organizes the world. Uexküll distinguishes two types of perceptual processes. The first is related to sensible experience, that records in memory the signals received in a sequence of actions in the real world (for example the memory of a trip into town), and are thus mediated by the senses. The second, however, is innate in the sense that the appearance of the world would be given in an almost "magical" way.... In the case of their Umwelt, animals can create imaginary worlds that Uexküll defines "magical", that allow a predictive behavior. (Berthoz, 2014)

Umwelt includes the world of things in the environment, the perceived world, the signals emitted from the subject is that by the objects, and actions that can be performed by each species. Above all, it includes the meaning of objects for each subject, to the extent that participate in the survival relationships and social relationships of the subject.

In describing the umwelt of the dog, Uexküll paints (literally) a room, whose chairs and whose dishes on the table are significant elements in the canine world, unlike textbooks that are wholly irrelevant. The dog has an idea in mind, an image search.

All the characteristics of the objects are in fact nothing other than the perceptive characters that are attributed to them by the subject with whom they have a relationship. (von Uexküll, in Berthoz, 2008)

The umwelt is, therefore, a dynamic, interactive concept that defines the relations between the physical world and living organisms, and constitutes the

Figure 1. The room in terms of functional tones connected with its object by a dog

Figure 2. The room in terms of functional tones connected with its object by man
Uexküll & von Uexküll, 1992.

basis and the assumption of intersubjectivity (Berthoz, 2009), an interface in which "the significance is conferred by the Act of the subject (von Uexküll & Müller, 2004). The subject builds up his world in accordance with his basic necessities and his tools of action, in a perspective which recalls Bergson and the phenomenological tradition attributable to Husserl and Merleau-Ponty, in the concept of enlacement proposed by Varela (Varela, Thompson, & Rosch, 1992), as in Von Foerster's second-order cybernetics idea, up to, on another plane, at the Mille Plateaux of Deleuze and Guattari (Deleuze & Guattari, 2003). In Uexküll as in Varela and Von Foerster it is possible to re-trace the roots of a bio-constructivism, which includes all living systems, psychological systems and social systems upstream of the different conceptualizations, and which has difficulty in establishing a common universe precisely because, When these authors write of "reality," they refer to not a universe, but a multiverse.

Alain Berthoz proposes a vision in which the subject navigates in his own umwelt led by a series of simplifying principles that optimize the process of perception - action and minimize the need for computation (Berthoz, 2008a).

These principles acknowledge to the subject the role of active constructor of the Umwelt other and define the Umwelt as an interface between an acting person and an "Umgebung" (entourage, environment, circle) that our Umwelt cannot embed (Merleau-Ponty, 2003)

In this perspective, the principles identified by simplifying it Berthoz are designed to cope with a measurable complexity in terms of data processing and decision between alternative opportunities, linking the meaning of complexity developed in the hard sciences.

Learning, as a fundamental strategy for facing complexity, is a peculiar adaptive process the living being, which develops in the umwelt. Cognitive skills can be considered as the result of evolutionary adaptations to an extremely small segment of the world as it is known to us today (Singer, 2009).

We should think about learning environments in terms of the students' umwelten, because these contain the structures that students perceive and act towards. It is these umwelten that change as students interact with their peers, teachers, and material structures. (Roth & Lawless, 2002)

SIMPLEXITY

Based on the work of Merleau-Ponty (1962), the world cannot be separated from the perceiver: we can know reality only in an experiential way.

The body's local situation is the birthplace of a singularity that is finally stabilized through a variation, where something or some event make sense defined by a subjective agent.

Faced with insurmountable difficulties posed by a multidimensional reality, the typical activities of living beings proceeds by reducing the dimensionality of the problems, a reduction that addresses the complexity simply acting, anticipating (Petit, 2012).

In action, in other words, the organism solves the complexity. According Berthoz, the brain solves the complexity of the outside world by producing perceptions consistent with the intentions regarding the future, memories of the past and the laws of the outside world that has internalized.

Basically, if our brain is forced to process all the information that the perception of the outside world gives him to be able to decide how to act in a certain situation, the task would be too costly and complex to allow an outlet for timely decision: we would constantly late and would not be able in the end to produce effective decisions. (Rivoltella, 2014)

REFERENCES

Berthoz, A. (1997). *Le sens du mouvement* (Vol. 223). Odile Jacob Paris.

Berthoz, A. (2008a). *Neurobiology of "Umwelt": How Living Beings Perceive the World*. Springer.

Berthoz, A. (2008b). The human brain "projects" upon the world, simplifying principles and rules for perception. In A. Berthoz (Ed.), *Neurobiology of "Umwelt"* (pp. 17–27). Berlin: Springer.

Berthoz, A. (2009). *Simplexité*. Editions Odile Jacob.

Berthoz, A. (2011a). Fondements cognitifs de la perception de l'espace. In J.-F. Augoyard (Ed.), Faire une ambiance = creating an atmosphere: actes du colloque international Grenoble 10-12 septembre 2008 (pp. 121-132). Grenoble: A la croisée.

Berthoz, A. (2014). *La vicarianza. Il nostro cervello creatore di mondi*. Torino: Codice.

Berthoz, A., & Petit, J.-L. (2006). *Phénoménologie et physiologie de l'action*. Odile Jacob.

Berthoz, A., & Petit, J. L. (2006). *Physiologie de l'action et Phénoménologie*. Paris: Odile Jacob.

Decety, J., & Jackson, P. L. (2004). The functional architecture of human empathy. *Behavioral and Cognitive Neuroscience Reviews*, *3*(2), 71–100. doi:10.1177/1534582304267187 PMID:15537986

Deleuze, G., & Guattari, F. (2003). *Millepiani*. Castelvecchi.

DiSalle, R. (2006). *Understanding Space-Time: The Philosophical Development of Physics from Newton to Einstein*. Cambridge, UK: Cambridge University Press. doi:10.1017/CBO9780511487361

Jeannerod, M. (1994). The representing brain: Neural correlates of motor intention and imagery. *Behavioral and Brain Sciences*, *17*(02), 187–202. doi:10.1017/S0140525X00034026

Kant, I. (1855). *Critique of Pure Reason*. London: Henry G. Bohn.

Leman, M. (2008). *Embodied Music: Cognition and Mediation Technology*. MIT Press.

Llinás, R. R. (2002). *I of the Vortex: From Neurons to Self.* Cambridge: MIT Press.

Merleau-Ponty, M. (1962). *Phenomenology of Perception: An Introduction.* Routledge & Kegan Paul.

Merleau-Ponty, M. (2003). *Nature: Course notes from the Collège de France.* Evanston, IL: Northwestern University Press.

Millican, P., & Clark, A. (1996). *Machines and thought.* Oxford, UK: Clarendon Press.

Petit, J. (2012). *Complexité-Simplexité - De la simplexité au champ phénoménal: La réponse du vivant à la complexité.* Retrieved from http://www.jeanluc-petit.com/sites/default/files/seminaire_complexite-simplexite_23-24.05.12.pdf

Prinz, W., & Hommel, B. (2002). *Common mechanisms in perception and action: Attention and performance XIX.* Oxford University Press.

Norton, J. (2004). The hole argument. In E. N. Zalta (Ed.), The Stanford Encyclopedia of Philosophy (Fall 2011 ed.). Stanford University.

Rivoltella, P. C. (2014). *La previsione. Neuroscienze, apprendimento, didattica.* La Scuola.

Roth, W. M., & Lawless, D. (2002). Scientific investigations, metaphorical gestures, and the emergence of abstract scientific concepts. *Learning and Instruction, 12*(3), 285–304. doi:10.1016/S0959-4752(01)00023-8

Singer, W. (2009). The Brain's View of the World Depends on What it has to Know. In A. Berthoz (Ed.), *Neurobiology of "Umwelt"* (pp. 39–52). Berlin: Springer. doi:10.1007/978-3-540-85897-3_5

Turing, A. (1947). *Lecture to the London Mathematical Society on 20 February 1947. In A. M. Turing's ACE Report of 1946 and other papers.* Cambridge, MA: MIT Press.

Varela, F. J., Thompson, E. T., & Rosch, E. (1992). *The Embodied Mind: Cognitive Science and Human Experience.* MIT Press.

von Uexküll, J. (1934). *Streifzüge durch die Umwelten von Tieren und Menschen: Ein Bilderbuch unsichtbarer Welten.* Berlin: Springer. doi:10.1007/978-3-642-98976-6

von Uexküll, J., & Müller, P. (2004). *Mondes animaux et monde humain: suivi de Théorie de la signification*. Paris: Pocket.

von Uexküll, J., & von Uexküll, T. (1992). *Jakob Von Uexküll's A Stroll Through the Worlds of Animals and Men*. Mouton de Gruyter.

ENDNOTES

1 Dans le traitement de l'espace, le problème pour le cerveau, c'est la multiplicité des espaces. Il n y a pas "l'espace", il y a une multiplicité formidable d'espaces. Deuxièmement, percevoir l'espace n'équivaut pas à percevoir la géométrie, mais à percevoir un mouvement. (Berthoz, 2011a)

2 Le cerveau n'est pas un ordinateur: il faut en tirer la conséquence en rejetant (au lieu de l'aménager superficiellement) une conception fonctionnaliste de la pensée en termes de computations symboliques qui relèverait d'une analyse purement logique indépendante de leurs réalisations biologiques. (Berthoz & Petit, 2006)

3 En effet, nous changeons en permanence de référentiels. J'utilise pour désigner cette action le terme de "manipulation de référentiels spatiaux", bien qu'il s'agisse d'une opération mentale et non manuelle. Selon moi, une telle richesse du répertoire de référentiels potentiels permet la vicariance fonctionnelle. (Berthoz, 2013)

Chapter 2
Space Representation in Children:
From Piaget to Neuroscience

ABSTRACT

With "Spatial Reference Frames" we refer to systems of coordinates by which the central nervous system encodes the relative positions of objects in space, including that of the body itself. A reference system is a way of representing the positions of the subjects / objects in space. The spatial position of an object can be represented in the brain with respect to different classes of reference points, which may be related or not to the position of the subject. In a nutshell, we can say that there are two types of transformations of space imagery: the allocentric spatial transformations, that involve a system of representation from object to object and encode information about the location of an object or its parts in relation to other objects, and egocentric spatial transformations that involve a system of subject-object representation. The human being switches from one code to another, depending on the contingent requirements, giving preference to one or another system according to a set of heterogeneous factors. The gender difference (male / female), for example, plays a key role. Even the individual cognitive strategies make use of different representations in a significantly different way. Manipulation of spatial reference systems constitute a "transnosographic trait" in various neurological and psychiatric disorders. Each of these diseases (autism, schizophrenia, epilepsy, spatial anxiety, Parkinson) reaches some of the structures involved in the manipulation of referential of different spaces. The chapter illustrates Piaget's study on the representation of space in the child and the use of different spatial coding systems, and provides a brief overview of the scientific debate following the Piagetian position.

DOI: 10.4018/978-1-5225-2455-7.ch002

SPATIAL FRAMES MANIPULATION

A key concept in the field of spatial processing concerns the definition of the reference systems used by the central nervous system to interpret sensory information and locate objects in space. With reference systems locution we refer here to coordinate systems through which the central nervous system encodes the relative positions of objects in space, including that of the body itself (Gaunet & Berthoz, 2000). In other words, a reference system is a way of representing the positions of the subjects / objects in space. The spatial position of an object can be represented in the brain with respect to different classes of reference points that can be related or independent from the position of the subject. In summary, we can say that there are two types of spatial image transformations: the allocentric spatial transformations, which involve system of representation from object to object and encode information about the position of an object or its parts in relation to other objects, and egocentric spatial transformations that involve a subject-object representation system.

The allocentric and egocentric spatial representations considerably differ. Spatial information provided by an allocentric representation are referred to an space external to perceiver; the information provided by an egocentric representation refers to a person who perceives with a defined orientation axis. In particular, the allocentric representation encodes the positions of the points in the space in the internal equivalent of a system of Cartesian or polar coordinates. The egocentric representation makes use of a special polar coordinate system whose origin is the ego (the perceiving subject) and the reference axis is the axis of orientation of the subject, by encoding the position of a point in terms of distance and angle to the subject.

In the allocentric reference system, the information relating to the position of an object are encoded according to the position of other objects. The location of an object is relative to the position of other objects. In the egocentric reference system, the information relating to the position of an object are encoded on the basis of body axis of the subject. The location of an object is relative to the position of the subject. You can represent the egocentric system using the Cartesian plane derived from crosses of the frontal plane and the sagittal plane.

The two types of representation coexist. The human being switches from one code to another, depending on the contingent requirements, giving preference to one or another system according to a number of heterogeneous factors. The gender difference (male / female), for example, plays a key role. Even

Table 1. Spatial reference frames: allocentric system

Information concerning the position of an object is encoded according to the position of other objects. The location of an object is relative to the position of other objects

Figure 1. Spatial reference frames: allocentric system

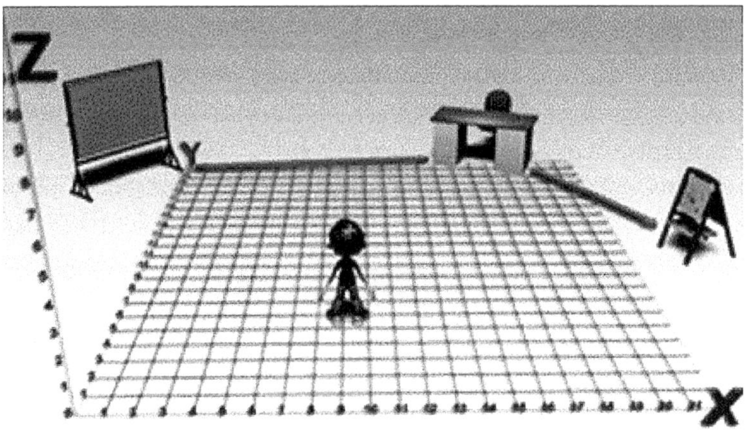

Table 2. Spatial reference frames: egocentric system

Information concerning the position of an object is encoded according to the body axis of the subject. The location of an object is relative to the position of the subject. In the graph below, the polar coordinates are used. You can represent the egocentric system using the Cartesian plane derived from crosses of the frontal plane and the sagittal plane

individual cognitive strategies make use of different representations. The transition from one code to another is related to the individual's evolutionary development (Berthoz, 2013).

Berthoz emphasizes as a coherent perception of one's body in relation to the environment is condition for use of the multiplicity of spatial references:

The necessary condition to use the repertoire of spatial reference systems of the ways equivalent actions, choosing one, or to address a problem simultaneously from different points of view, is to have a coherent sense, unique, stable of our body and its relationship with the environment. (Berthoz, 2013)

Figure 2. Spatial reference frames: egocentric system

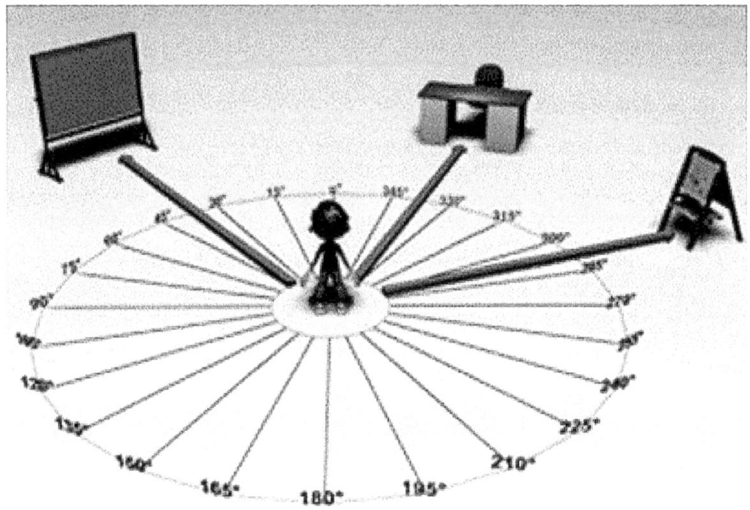

The concept of space, in this context, must be understood by abandoning Euclidean perspective and by adopting the idea of space as a potential of action: "the space is divided into zones corresponding to actions that can be done" (Berthoz, 2013).

This idea of space boasts illustrious ancestors:

Locate an object in space simply it means to represent the movements that would be needed to achieve it. This is not to represent the same movements, but simply the muscular sensations that accompany them. (Poincaré, 1895)

According to Poincare, simply, the plurality of reference frames and la coordination qui en résulte qui est l'espace. For Poincare movement is the foundation of the concept of building space and offers us the opportunity to change our interpretation of the properties of the world.

Einstein, in this regard, explicitly quotes Poincare:

Poincaré is right. The fatal mistake that mental necessity that precedes any experience is the basis of Euclidean geometry is due to the fact that the empirical basis underpinning the axiomatic construction of Euclidean geometry has been forgotten. The geometry must be considered as a physical science, the utility should be judged by its relationship with the sensory experience. (Einstein, 1990)

Neuroscience research has also shown that, in this context, it is more appropriate to speak of spaces, and not of space. In fact, different neural networks are involved in managing the various spaces: the space of the body, the space of prehension, the surrounding environment space, the distant environment space. In other words, to take the cup of coffee on the table we use neural networks and geometries different from those we use to imagine a path between two neighborhoods of a city. The different networks prefer different encodings, as well in the parietal cortex encoding is egocentric, and refers to the observer's body, while in the hippocampus encoding is allocentric. Little is known about the mechanisms that the brain uses for transcoding of different representations, to build a coherent synthetic representation and then enable change of perspective. Of course, different combinations of codes, spaces and underlying neural networks give rise to different cognitive strategies.

PIAGET'S THREE MOUNTAINS

The original ideas of Piaget on mental development have focused on egocentrism in early childhood, based on experimental studies, such as the famous *three mountains task* (Piaget & Inhelder, 1948).In this test, a child must indicate the point of view of an observer who is in a different position.

Trisciuzzi (2011) summarizes:

Figure 3. Piaget's three mountains task

A well-known experimental evidence is that offered by Piaget: starting from a drawing that represents a person who is photographing three pyramids or piles (sand or other), he asks the child: "How will be the photograph? Will be a, b, c, d, or e?"

Bearing in mind the position of the photographer, the choice should fall on the figure 'c'. This is the position of those who are able to place themselves in spatial situations other than those watching only centrally; a child, however, still immersed in a egocentric point of view, is difficult to choose a figure out of its narrow point of view. Going out from self, therefore, means being able to see the reality (not only spatial, but also a mental reality, linguistic or otherwise) from several points of view, however, other than their own. (Trisciuzzi & Zappaterra, 2011)

Using this paradigm, children up to 7 years, seem not to have the ability to evaluate a point of view different from their own.

Only when they reach the stage of concrete operations, between 7 and 12 years, acquire the "decentralization skills." This allows them to take into account multiple aspects of a task to solve it. According to Piaget, the egocentrism, defined as inability to decentralize and take the perspective of another person, is the norm in young children.

In fact, the age between 6-7 years and 12-13 years is defined, in common parlance, a "critical period." It is the age when the child learns the change

Figure 4. Three mountains task
Trisciuzzi & Zappaterra, 2011.

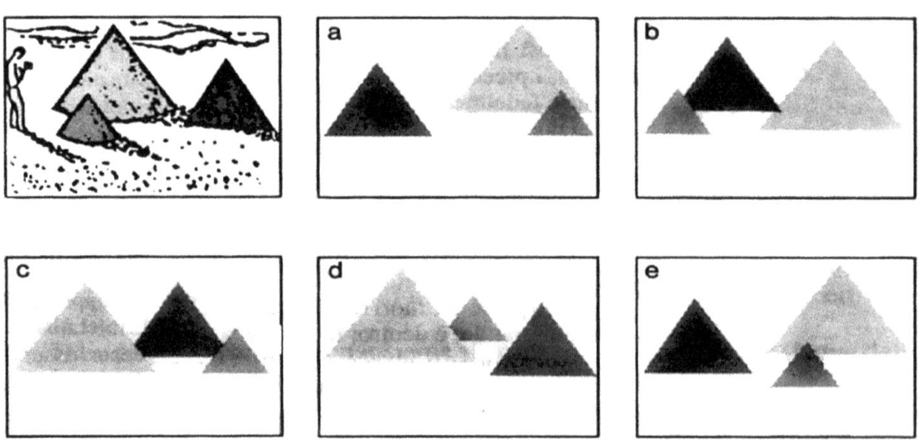

point of view, realizes that the world cannot be seen one way, that space can be manipulated, that it is possible to consider the thoughts and emotions of others. It is the age in which the capacity for empathy develops, which is not only the emotional contagion that takes place between mother and child, but it is the ability to remain yourself putting while seeing the world through the others' eyes.

The ability to change point of view is extremely important from the cognitive point of view: if, during the critical period when you open a "window" for this power, this is not acquired, once you close the "window", the child will remain locked in a unique vision of the other, which will be a source of sectarianism and hatred.

We can imagine the child of seven to ten years, blocked in its ability to develop different cognitive strategies, such as a person trapped in a maze with a single output, a single view of the world. To leave the path traced, the child has to make a decentralization operation, switch between " egocentric perception "to an " allocentric perception ", finding a "shortcut", by inhibiting the usual way (Berthoz & Jorland, 2004).

Incidentally, the passage from the egocentric perception to a "decentralized perception" is one of the nodal points of criticism that Vygotski (Vygotsky & COLE, 1978) moves to Piaget, contesting a "linear" conception of development (Tryphon & Vonèche, 1996).

Piaget's criticisms have been numerous, and we will give a quick review in the next paragraph. It should now be noted that many of these criticisms target the tools used (or, better, the designed tasks), and often the critical researchers with Piaget achieve quite different results. However, criticisms are about how Piaget conducted experiments, but they do not concern the conceptual framework that they often share.

SCIENTIFIC DEBATE ABOUT PIAGET'S THEORY

Piaget's theory on 'egocentrism has provoked a lively debate, of which Perner offers a comprehensive overview (Perner, 1991).

Martin Hughes (1979) has argued that one of the main limitations of the "three mountains" is the lack of interest that the experiment arouses in children. As a result, Hughes has proposed a variant that, while sharing in fact the Piagetian conceptual framework, differs in the practical implementation.

Hughes:

Devised a task which made sense to the child. He showed children a model comprising two intersecting walls, a 'boy' doll and a 'policeman' doll. He then placed the policeman doll in various positions and asked the child to hide the boy doll from the policeman"…. Hughes argues, on the basis of experimental results that, since the four years, the children are able to take the view of another. (Hughes & Donaldson, 1979)

Rochat (1995) showed that children 3 years of age are able to discriminate what they can directly reach by what, instead, can be reached by someone else. The conclusion of Rochat is clear: since the three years the children can take another person's perspective (perspective-taking), they are capable of spatial decentralization and flexibility in the spatial reference systems depending on the operation to be performed (Rochat, 1995).

The point that seems particularly interesting for the purposes of discourse that here is being carried out, is not the disagreement between Piaget and Rochat about the age at which the child can take the perspective of others. Beyond the age parameter, both share the same definition of "allocentric" and "egocentric". The task of the three mountains requires to take another person's visuospatial perspective, a perspective which, although of a different subject, however, remains an egocentric perspective.

Figure 5. Policeman task
Hughes & Donaldson, 1979.

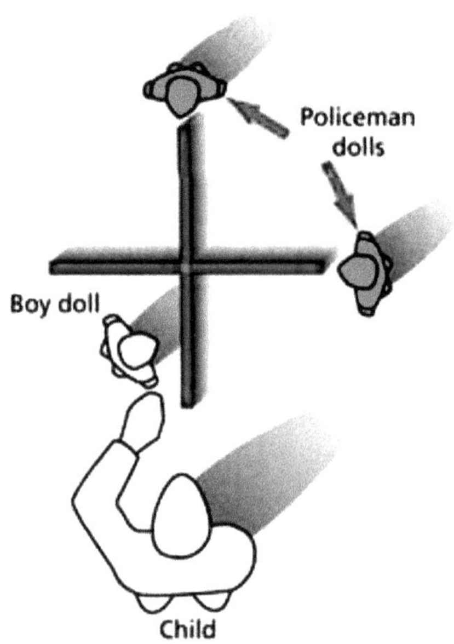

The task of the three mountains, according to Frith and de Vignemont, is always based on an egocentric representation of the object and cannot inform us on the capacity to take an allocentric perspective by younger children (Frith & De Vignemont, 2005).

This appears to agree with the thesis of Vogeley and Fink, according to which:

The difference between the first-person perspective and third-person perspective is that the third person requires a translocation of egocentric point of view. (Vogeley & Fink, 2003)

The ability to assume an allocentric perspective, however, is not reducible to the mechanical assumption of another's position in space. The central node is the ability to make a:

Mental rotation on themselves, in relation to the environment or to an object of the environment, maintaining the main environmental perspective in question. (Berthoz, 2011b)

REFERENCES

Berthoz, A. (2011b). *La semplessità*. Torino: Codice.

Berthoz, A. (2013). *Vicariance (La): Le cerveau créateur de mondes*. Odile Jacob.

Berthoz, A., & Jorland, G. (2004). *Empathie (L')*. Paris: Editions Odile Jacob.

Einstein, A. (1990). *Conceptions scientifiques*. Paris: Flammarion.

Frith, U., & De Vignemont, F. (2005). Egocentrism, allocentrism, and Asperger syndrome. *Consciousness and Cognition*, *14*(4), 719–738. doi:10.1016/j.concog.2005.04.006 PMID:15996486

Gaunet, F., & Berthoz, A. (2000). Mental rotation for spatial environment recognition. *Brain Research. Cognitive Brain Research*, *9*(1), 91–102. doi:10.1016/S0926-6410(99)00038-5 PMID:10666561

Hughes, M., & Donaldson, M. (1979). The use of hiding games for studying the coordination of viewpoints. *Educational Review*, *31*(2), 133–140. doi:10.1080/0013191790310207

Perner, J. (1991). *Understanding the representational mind.* Cambridge, MA: The MIT Press.

Piaget, J., & Inhelder, B. (1948). *La représentation de l'espace chez l'enfant.* Paris: Presses Universitaires De France.

Poincaré, H. (1895). L'espace et la géométrie. *Revue de Metaphysique et de Morale*, *3*(6), 631–646.

Rochat, P. (1995). Perceived reachability for self and for others by 3-to 5-year-old children and adults. *Journal of Experimental Child Psychology*, *59*(2), 317–333. doi:10.1006/jecp.1995.1014 PMID:7722438

Trisciuzzi, L., & Zappaterra, T. (2011). Dislessia, disgrafia e didattica inclusiva. *Annali Della Pubblica Istruzione*, *2*, 51–76.

Tryphon, A., & Vonèche, J. J. (1996). *Piaget-Vygotsky: The Social Genesis of Thought*. Psychology Press.

Vygotsky, L. S., & Cole, M. E. (1978). *Mind in Society: The Development of Higher Psychological Processes*. Harvard University Press.

Chapter 3
Space Representation and Gender Differences

ABSTRACT

Scientific literature highlighted gender differences in spatial orientation. In particular, men and women differ in terms of the navigational processes they use in daily life. Scientific literature highlighted that women use analytical strategies while men tend to use holistic strategies. According to classical studies, males show a net advantage at least in the two categories of mental rotation and spatial perception. Subsequently, brain-imaging studies have shown a difference between males and females in the activity of brain regions involved in spatial cognition tasks. What we can say with certainty is that, given the complex nature of the subprocesses involved in what we call spatial cognition, the gender differences recorded by numerous scientific studies conducted in this field are closely related to specific measured abilities. The evidence that emerges with certainty from diverse studies is, however, that of a huge variety of strategies that differ according to sex, context, purpose to reach, education, age, and profession. In the study presented here, the gender and age-related tests show a significant sex-based difference perspective-taking tasks, but there is no gender-based difference in the mental rotation task.

DOI: 10.4018/978-1-5225-2455-7.ch003

GENDER DIFFERENCES: A POPULAR TOPIC

Numerous data in the literature provide evidence for gender differences in spatial orientation. (Lambrey & Berthoz, 2007)

Gender differences have been a popular topic in social sciences for many years.

Regarding the specific field of spatial cognition, it should be specified that both common belief and scientific literature claim that men and women differ in spatial abilities.

The 1974 seminal study by Maccoby and Jacklin, in which spatial cognition was considered a unitary capacity in which males excelled in comparison to females, marrying well with common beliefs, gave rise to the commonplace of a supposed male greater ability which is feeding itself by metabolizing results from specific studies.

What we can say with certainty is that, given the complex nature of the subprocesses involved in what we call spatial cognition, the gender differences recorded by numerous scientific studies conducted in this field are closely related to specific measured abilities. Moreover, given once again the peculiar nature of the field of study, the choice of instruments used for measurements is not at all neutral. Specifically, the differences relate to paper-and-pencil test based studies compared to those that take into account the ecological variable and are based on tasks performed in real or virtual spaces

A 1985 study by Linn and Petersen (Linn & Petersen, 1985) classified spatial tests, distinguishing the three categories of mental rotation, spatial perception, and spatial visualization. According to classical studies (Benton, Varney, & deS Hamsher, 1978; Inhelder & Piaget, 1958; Shepard & Metzler, 1971) quoted in de Goede (de Goede, 2009), males show a net advantage at least in the two categories of mental rotation and spatial perception. Subsequently, brain-imaging studies have shown a difference between males and females in the activity of brain regions involved in spatial cognition tasks.

A brain imaging study by Grön, Wunderlich, Spitzer, Tomczak, and Riepe (2000) have shown that males and females recruit different brain regions in navigation. They found that men primarily engaged the left hippocampal region. In women on the other hand the right parietal and right prefrontal areas were activated. These differences might reflect typical gender differences in navigational approaches, that is a strong reliance in females on landmarks and the usage of both landmarks and geometry in males, which would in general provide males with an advantage in navigational tasks (de Goede, 2009).

Other studies have addressed the cognitive dimension of gender differences in spatial cognition, suggesting that these differences are related to coding strategies rather than differences in specific abilities.

Schmitz (1997) has suggested that different strategic preferences already develop in childhood; since boys, as compared to girls, tend to have a larger range of spatial experience, they would experience less spatial anxiety later on in live. Higher levels of anxiety would induce a stronger dependency on landmarks, which provides an explanation for the preference for this cue type in females. (de Goede, 2009)

NATURE OR NURTURE

The biological structure is fundamentally important, but its links with spatial perception are not of a mechanical nature. Studies involving gender differences have largely debated the link between the right and left hemispheres of the brain with linguistic and spatial functions in men and women. Clinical trials on patients with right hemisphere have not demonstrated, However, a more devastating effect of such damage to the spatial performance of men than women, as it would have been easy to suppose in accordance with the assumption made earlier. The results of a Kimura study suggest that normal differences between men and women on line rotation and orientation activities should not be the result of different degrees of dependence on the right hemisphere.

My laboratory has studied the ability of patients with damage to one hemisphere of the brain to visualize the rotation of certain objects. As expected, for both sexes, those with damage to the right hemisphere got lower scores on these tests than those with damage to the left hemisphere did. Also, as anticipated, women did not do as well as men on this test. Damage to the right hemisphere, however, had no greater effect on men than on women. (Kimura, 1992)

According to Kimura, some other brain systems may be mediating the higher performance by men.

Berthoz summarizes:

There are anatomical and neuroendocrinal bases for gender differences..., but it is now accepted that, whatever the origin (nature or nurture) of these differences, men and women do not process spatial information, for instance, in the same way.

According to Berthoz, women are more dependent on visual field for spatial orientation. They are statistically better in cognitive tasks that can be verbally mediated and in storing significant forms that can be called to recognize. Men and women do not react the same way to a sensory conflict. These differences are not due to education, although it can of course have a great influence. They originate in anatomical differences in the brain, but also hormonal factors - estrogen changes for women and testosterone for men (Berthoz, 2013).

For travel memory, women tend to adopt more egocentric and sequential strategies. Berthoz suggests that this is related to their preference for what can be verbally mediated. In other words, women prefer a verbal description of travel, which is of a sequential nature. This is probably due to the cerebral lateralization process.

Men, on the other hand, prefer more allocentrics strategies. In general, males are statistically more efficient in mental rotation tasks, such as changing perspective by reading a map. Some studies point out that the mental representation of large environments in males contains more "metric" information in men than in women, while females relied more on information about the actual references in the environment during navigation.

On the whole, variation between men and women tends to be smaller than deviations within each sex, but very large differences between the groups do exist—in men's high level of visualspatial targeting ability, for one. (Kimura, 1992)

The evidence that emerges with certainty from diverse studies is, however, that of a huge variety of strategies that differ according to sex, context, purpose to reach, education, age, and profession.

Berthoz speaks, in this regard, of functional vicariance.

The concept of vicariance and the links between vicariance and spatial navigation strategies will be discussed in Chapter 6.

MALES, FEMALES, BOYS, AND GIRLS

Although it used to be thought that sex differences in problem solving did not appear until puberty, the accumulated evidence now suggests that some cognitive and skill differences are present much earlier. For example, researchers have found that three- and four-year-old boys were better at targeting and at mentally rotating figures within a clock face than girls of the same age were. (Kimura, 1992)

In the foregoing study, the gender and age-related tests show a significant sex-based difference perspective-taking tasks, but there is no gender-based difference in the mental rotation task. Success rates for three tasks seem to indicate a progressive improvement in performance in the three tasks in relation to increasing age for both sexes.

REFERENCES

Benton, A. L., & Varney, N. R. (1978). Visuospatial judgment: A clinical test. *Archives of Neurology*, *35*(6), 364–367. doi:10.1001/archneur.1978.00500300038006 PMID:655909

de Goede, M. (2009). *Gender differences in spatial cognition.* Utrecht University.

Grön, G., Wunderlich, A. P., Spitzer, M., Tomczak, R., & Riepe, M. W. (2000). Brain activation during human navigation: Gender-different neural networks as substrate of performance. *Nature Neuroscience*, *3*(4), 404–408.

Inhelder, B., & Piaget, J. (1958). *The Growth of Logical Thinking From Childhood to Adolescence: An Essay on the Construction of Formal Operational Structures (Developmental Psychology).* Basic Books. doi:10.1037/10034-000

Kimura, D. (1992). Sex differences in the brain. *Scientific American*, *267*(3), 118–125. doi:10.1038/scientificamerican0992-118 PMID:1298222

Lambrey, S., & Berthoz, A. (2007). Gender differences in the use of external landmarks versus spatial representations updated by self-motion. *Journal of Integrative Neuroscience*, *6*(03), 379–401. doi:10.1142/S021963520700157X PMID:17933018

Linn, M. C., & Petersen, A. C. (1985). Emergence and characterization of sex differences in spatial ability: A meta-analysis. *Child Development*, *56*(6), 1479–1498. doi:10.2307/1130467 PMID:4075870

Schmitz, S. (1997). Gender-related strategies in environmental development: Effects of anxiety on wayfinding in and representation of a three-dimensional maze. *Journal of Environmental Psychology*, *17*(3), 215–228. doi:10.1006/jevp.1997.0056

Shepard, R. N., & Metzler, J. (1971). Mental rotation of three-dimensional objects. *Science*, *171*(3972), 701–703. doi:10.1126/science.171.3972.701 PMID:5540314

Chapter 4
Space and Media:
Digital Umwelten

ABSTRACT

The tool identified for data collection of this research project is a video game, which makes the topic of the representation of space in videogame an absolutely relevant aspect for the project. This work bases on the statement of Jenkins, according to which "game space never exists in abstract, but always experientially". In the current generation of video games, talking about position of the camera assumes a different value than in film or television language, assuming the meaning of point of view from which the game is visually (and auditory) presented and determines the spatial perspective of a computer game. The most common distinction, with respect to the position of the camera, is between First Person Camera, where space is presented from the perceptive perspective of the player's avatar and Third Person Camera, where the perspective is not directly the one of the avatar. This category, in fact, is very extensive, and poorly lends itself to a single definition. Under the umbrella of Third Person Camera are both perspectives associated with the avatar, but framing it externally (a camera follows the avatar) and those in which the camera is fixed. Moreover, the position of the camera compared to the avatar (from behind, left, right, Orbit Camera, etc.), or with respect to the environment (from above, from a precise point of reference) is not a neutral choice. In the present work, we use the categorization proposed by Britta Neitzel (Neitzel, 2002), which, taking up the work of Jean Mitry about The Aesthetics and Psychology of the Cinema (Mitry & King, 1997), distinguishes between subjective, semisubjective or objectives views. The chapter provides examples of different perspectives, and introduces the concept of Natural User Interfaces, which include movements based on input and output, on discretion, on voice, and evolve towards an efficient use of the senses in the interaction with machines.

DOI: 10.4018/978-1-5225-2455-7.ch004

STORYWORLD, WORLDSPACE, GAMESPACE

The tool identified for data collection of this research project is a video game, which makes the topic of the representation of space in videogame an absolutely relevant aspect for the project.

The opportunity to use digital tools in this particular field of investigation has experimentally established by Berthoz:

We designed a behavioral study in which participants interacted spontaneously with a life-sized virtual tightrope walker walking forward, backward and leaning to her left or right on a rope. Here, we report results showing that participants automatically embodied the avatar's leaning movements. Moreover, the form of the participants' motor behavior (i.e., automatic leaning movements to their right and left when the tightrope walker was leaning to her own right and left, respectively) revealed that participants, using mental imagery, located spontaneously themselves in the avatar body position, suggesting that they embodied the avatar's visuo-spatial perspective in the avatar's body position to take its visuo-spatial perspective, suggesting that embodiment process is not necessarily exhibited by a physically mirroring body posture. We further propose a model of self–other interaction showing how perspective-taking mechanisms may relate on mental body transformation and enabling to deepen the description of the different sorts of inter-subjectivity. (Berthoz & Thirioux, 2010)

Specifically in the context of video games, should be put to the premise of the relationship between space, perspective and virtual environment the consideration that many video games take the first-person perspective as the dominant perspective (FPC - First Person Camera) but at the same time allow to change perspective if necessary (TPC - Thirdh Person Camera), and the habit to switch between the different perspectives is established and considered effective among gamers.

This leads immediately to consider the peculiarities of the idea of space in videogames.

The experience of any space in video games varies depending on the player's presuppositions regarding the forms and limits of game space, the importance and use of a particular space to the game narrative, and the player's vantage point in the space. Once the player learns the conventions of both the geometric and experiential aspects of video game space, the player can then begin to both play and dwell within these spaces. (Taylor, 2002)

The idea of the game space as an experiential space is very well stated by Jenkins:

When gamer magazines want to describe the experience of gameplay, they are more likely to reproduce maps of the game world than to recount their narratives. As I have noted elsewhere, these maps take a distinctive form - not objective or abstract top-down views but composites of screenshots which represent the game world as we will encounter it in our travels through its space. Game space never exists in abstract, but always experientially. (Jenkins, 2007)

Therefore, we need first of all distinguish the space experienced by the user from the spatial representation of the environment of the game as a whole. The perspective adopted here is the one suggested by Jesper Juul (Juul, 2005), of which Thon presents a summary:

There is a wide variety of ways in which computer games can construct the space in which they take place, from "all text-based") via various forms of two-dimensional spaces to "interactive three-dimensional environments".... Many if not most contemporary computer games present a three -dimensional space on a two-dimensional screen.... Before we can examine more closely the various forms of spatial perspective that can be found in such games, it has to be made clear to which parts of these games we refer. Since many computer games are set in complex fictional worlds, one has to distinguish between the space of the fictional world as a whole and the spaces that the player can interact with through the interface. Jesper Juul draws a similar distinction between "world space" and "game space". Since most of the events in computer games take place in the game space, it seems to be mainly this part of the space of the fictional world that is of interest with regard to the question of spatial perspective in computer games. (Thon, 2009)

The experiential space of video games then coincides with the gamespace, as stated by Juul. Conventionally, the benchmark used by players to distinguish the possible perspectives is the position of the camera (First Person Camera, Thirdh Person Camera, etc.). This is not surprising; according to Wolf:

Most games representing their diegetic space as an interactive three-dimensional environment follow, to some degree, the precedent set by the space represented in the classical Hollywood film. Spaces and the objects in them can be viewed from multiple angles and viewpoints which are all linked together in such a way as to make the diegetic world appear to have at least enough spatial consistency so as to be navigable by the player. The extent to which the player is allowed to freely navigate the space varies widely. (Wolf & Baer, 2010)

Quite simply, the fundamental difference, compared to Hollywood's perspective identified by Wolf, is given by the interaction:

Unlike the spaces of film, paintings, and photography, videogame spaces are spaces that are both observed and engaged directly; they are thus experiential spaces. (Taylor, 2002)

In fact, in the current generation of video games, talking about position of the camera assumes a different value than in film or television language, assuming the meaning of point of view from which the game is visually (and auditory) presented and determines the spatial perspective of a computer game.

The most common distinction, with respect to the position of the camera, is between First Person Camera, where space is presented from the perceptive perspective of the player's avatar and Third Person Camera, where the perspective is not directly the one of the avatar.

This category, in fact, is very extensive, and poorly lends itself to a single definition. Under the umbrella of Third Person Camera are both perspectives associated with the avatar, but framing it externally (a camera follows the avatar) and those in which the camera is fixed. Moreover, the position of the camera compared to the avatar (from behind, left, right, Orbit Camera, etc.), or with respect to the environment (from above, from a precise point of reference) is not a neutral choice. Therefore we prefer to use the categorization proposed by Britta Neitzel (Neitzel, 2002), which, taking up the work of Jean Mitry about The Aesthetics and Psychology of the Cinema (Mitry & King, 1997), distinguishes between subjective, semisubjective or objectives views.

Points of View

Based on this categorization, a point of view is subjective when the position from which the gaming space is presented coincides with the position of the avatar (FPC - First Person Camera. The players, therefore, cannot see the body of the avatar, as they are able to see the weapons or the avatar's hands. This perspective is also frequently used to represent the point of view of a driver in a vehicle. in the FPC games is common to make use of positional audio, in which the volume of the environmental sound varies depending on the position of the sound source with respect to the avatar (Rollings & Adams, 2006).

The semisubjective perspective is given when the point of view is connected to the avatar's movements. In Neitzel words:

Point of view is connected to the movements of the avatar; it is not a substitute for the viewpoint as in case of the subjective POV, butrather a viewing-with. (Neitzel, 2002)

Thon emphasizes:

Figure 1. Screenshot from the computer game Stalker: Shadow of Chernobyl

Although the spatial position of the avatar is not the same as that of the camera, the camera's position is always linked to the avatar. (Thon, 2009)

Finally, when the perspective is not connected to location and movements of the avatar, it is possible to talk of objective point of view:

The objective point of view shows a game space from a position that is not part of this game space (as is the case with a subjective point of view) and is not connected to an entity in the game space (as is the case with a semisubjective point of view). (Thon, 2009)

Tracking Camera

After having distinguished the different points of view, it should be emphasized that most of today's games not only combines the different types of perspective, but gives players the opportunity to control the camera and select the point of view considered more functional to the current situation:

Many contemporary games not only combine various forms of spatial perspective but also allow their players to control camera movements (which is an essential part of the gameplay in most strategy games) and switch

Figure 2. Semisubjective perspective in Resident Evil 6. Camera follows avatar.

Figure 3. Avatar-independent perspective: Command & Conquer red alert

between different perspectives themselves.... It can be concluded that many contemporary computer games allow their players an ever greater amount of control over the spatial perspective(s) used in the presentation of the game space. Finally, it may be noted that while players generally like the opportunity to take control of the camera, they rarely use the possibility to change the default point of view. This has to do with the fact that the default point of view is often best suited to the interaction with the game space required by the game. (Thon, 2009)

NATURAL USER INTERFACES

The massive spread of natural interfaces in videogames (based on devices such as, for example, the wiimote or Kinect) has helped to redefine the concept of space, making "sensitive" the gamer space (Di Tore, Aiello, Di Tore, & Sibilio, 2012).

Natural interfaces (NUI - Natural User Interface) and gesture recognition technologies (Gesture Recognition), reproduce, in the digital environment,

Figure 4. Resident Evil 2 camerawork
Rollings & Adams, 2006; Licensed under Fair use via Wikipedia -http://en.wikipedia.org/wiki/File:Resident-evil-2-camerawork.jpg#/media/File:Resident-evil-2-camerawork.jpg

the fundamental questions of phenomenology, confirming how the actions incorporated within a digital interface are "fluid passages and functional between physical and virtual environments" (Hansen, 2006). The Open Source community NuiGroup, active since 2006 in the creation and sharing of standards and technical innovation for human machine interaction, defines natural interfaces:

Natural User Interface (NUI) is the next metaphysical paradigm shift in man machine interaction (MMI) also known as human computer interaction (HCI). Beginning with the Command Line Interface (CLI) and followed by the Graphical User Interface (GUI), we are now in the midst of discovering the next phase of a more organic interfaces which are based on more traditional human interaction paradigms such as touch, vision, speech and most importantly creativity. (NUIGroup, 2009)

Natural interfaces include movements based on input and output, on discretion, on voice, and evolve towards an efficient use of the senses in the interaction with machines.

For example, Microsoft Kinect, an accessory of the Xbox, can detect the segments of the body of a user placed in front of the camera and use this information to move an avatar in a video game.

In fact we are witnessing, not in the MIT labs, but in the bedrooms of middle school students, to forms of HCI (Human Computer Interaction) that go beyond the flattening on the Cartesian plane that has characterized so far the video game phenomenon.

The analysis of the gesture-based technology potential is developed from the knowledge that the devices that encourage touching, moving and exploring are considered basically interesting for education and training, especially within a vision that aims to enactive knowledge as a process that requires the participation of the brain, body and environment.

The gesture-based computing probably will allow to take control of the educational experience through the body and the voice, to transform the educational act in a natural and interactive experience, and to interact with a virtual environment and to "manipulate", influence and turn it into an intuitive manner. Knowledge becomes embodied in the subjects who learn and in the environments that synergistically include them (Bechara, Damasio, Tranel, & Damasio, 1997) converging more and more towards the idea that the body can be considered as one of the structuring dimensions in formation of the architecture of knowledge and thought.

REFERENCES

Bechara, A., Damasio, H., Tranel, D., & Damasio, A. R. (1997). Deciding advantageously before knowing the advantageous strategy. *Science*, *275*(5304), 1293–1295. doi:10.1126/science.275.5304.1293 PMID:9036851

Berthoz, A., & Thirioux, B. (2010). A Spatial and Perspective Change Theory of the Difference Between Sympathy and Empathy. *Paragrana*, *19*(1), 32–61. doi:10.1524/para.2010.0003

Di Tore, S., Aiello, P., Di Tore, P. A., & Sibilio, M. (2012). New Forms of Human Machine Interaction. *Academic Journal of Interdisciplinary Studies*, 165.

Hansen, M. B. N. (2006). *Bodies in Code: Interfaces with Digital Media.* Routledge.

Jenkins, H. (2007). Transmedia storytelling 101. *Confessions of an Aca/Fan: The Official Weblog of Henry Jenkins.*

Juul, J. (2005). *Half-real: Video games between real rules and fictional worlds.* The MIT Press.

Mitry, J., & King, C. (1997). *The Aesthetics and Psychology of the Cinema.* Indiana University Press.

Neitzel, B. (2002). *Point of View and Point of Action. A Perspective on Perspective in Computer Games.* Paper presented at the Challenge of Computer Games Conference, Lodz, Poland.

NUIGroup. (2009, May). *Multi-Touch Technologies.* Community Release. Author.

Rollings, A., & Adams, E. (2006). Fundamentals of game design: New Challenges for Character-Based AI for Games. In Artificial Life and Puzzle Games. Prentice Hall.

Taylor, L. N. (2002). *Video games: Perspective, point-of-view, and immersion.* University of Florida.

Thon, J.-N. (2009). Perspective in Contemporary Computer Games. *Point of View, Perspective, and Focalization: Modeling Mediation in Narrative,* 279-300.

Wolf, M. J. P., & Baer, R. H. (2010). *The Medium of the Video Game.* University of Texas Press.

Chapter 5
IkeWYSe:
I Know What You See

ABSTRACT

The first product of the project presented here was a prototype of video games, named IKeWYSE, which constituted the data collection tool for the experimental research project. The videogame prototype realized requires the user to navigate in a three-dimensional space through an avatar. User deals with three different tasks, two of which are designed to measure the skills of perspective taking, while the third task is calibrated on the ability of mental rotation. The default point of view is a semi-subjective view with the camera following the avatar. The player has the option to select other views, going through semi-subjective, subjective and objective point of view. The game was tested between January and March 2015, with a group of 70 children, 35 males and 35 females, from the third, fourth and fifth primary school class and from a first secondary school class of the Istituto Comprensivo San Valentino Torio, in the province of Salerno. Each user performs 10 attempts for each task. The software records the beginning of each game, the user data (age and gender), and, during the game, time for each attempt and the result (success / failure) of the attempt. The present chapter presents the research design, the path for design and development of the videogame, the methods for data collection and discusses the results obtained.

DOI: 10.4018/978-1-5225-2455-7.ch005

INTRODUCTION

The scientific debate on the relationship between perceptive egocentrism and perspective taking has focused on two main factors. First, the age when the child leaves perceptive egocentrism. The disagreement between the various positions is strong, and results in sometimes very distant evaluations. If Piaget suggests a "window" of time between 6 and 11 years of age, other scholars, based on different experimental trials, have supported different thesis, moving the threshold to five, four, and even three years (Rochat, 1995).

In addition, the ability of perspective-taking, taken individually, does not seem sufficient to demonstrate the ability, in the child, to have a coherent representation of the space, to allow the manipulation of viewpoints. Berthoz, in this regard, relates the abandonment of perceptive egocentrism with a more complex mechanism than the perspective-taking.

This mechanism consists in the possibility of performing a mental rotation on ourselves by maintaining a main perspective of the environment (Berthoz, 2011b).

According to Berthoz, leaving perceptive egocentrism resides in the ability, based on the mental rotation skills, to simultaneously use egocentric, allocentric and heterocentric perspectives.

Secondly the gender difference plays a key role in spatial thinking. Berthoz summarizes:

Numerous data in the literature provide evidence for gender differences in spatial orientation. In particular, it has been suggested that spatial representations of large-scale environments are more accurate in terms of metric information in men than in women but are richer in landmark information in women than in men. (Lambrey & Berthoz, 2007)

Criticisms of Piaget from Hughes and Rochat demonstrate the difficulty in conceiving paradigms of research on the topic. In this sense, the evolution of digital systems for the representation of space provides valuable tools for building effective instruments.

Starting from this assumption, the present research project intended to investigate the relationship between age, perspective taking skills and mental rotation, and the relationship between gender difference, perspective taking skills and mental rotation.

The hypothesis about the relationship between age and skills under investigation, gained in the study of the scientific literature, is that these skills are developed in a very extensive range, from 4/5 to 11/12 years of age, and covers, in fact, much of the first cycle of education.

In particular, if the skills of perspective taking begin to occur early as three years old, the ability of mental rotation, that allows to integrate the different perspectives into a coherent functional representation of space, fully manifests itself only at the turn of the ten / eleven years, confirming the view taken by Piaget and Inhelder in 1948 (Piaget & Inhelder, 1948).

As for the difference between the genders, the assumption is that the different behavior between genders is not a variable that intervenes only during the acquisition of the skills being investigated, but that represents a structural difference, and persists into adulthood.

The project included the development from scratch of a non-invasive research tool (a videogame), designed to be meaningful to the target identified - that could make sense to the children, in the expression of Hughes (Hughes & Donaldson, 1979) - and to take advantage of the full potential of space representation offered by the new media and of the confidence that the current generation of primary school students shows to possess with such systems.

VIDEOGAME DESIGN AND DEVELOPMENT

The videogame prototype realized requires the user to navigate in a three-dimensional space through an avatar. User deals with three different tasks, two of which are designed to measure the skills of perspective taking, while the third task is calibrated on the ability of mental rotation. The default point of view is a semi-subjective view with the camera following the avatar. The player has the option to select other views, going through semi-subjective, subjective and objective point of view.

The use of terminology "subjective point of view", "semi-subjective point of view", objective point of view" (see Figure 1 of Chapter 2) refers to the classification proposed by J. Mitry in *The Aesthetics and Psychology of the Cinema* (Mitry & King, 1997).

In the first task, the avatar of the player is located in a park and has in front of him two men. A window in overlay shows the point of view of one of the two men. The player's task is to indicate which of the two men belong the views shown in the overlay window (Figure 1 - Screenshot of the first task.)

The position of the avatar is fixed (the user can change your point of view, but cannot move the avatar in space), while the position of the elements on the scene is random, according to a predetermined pattern. At each new attempt,

Figure 1. First task screenshot

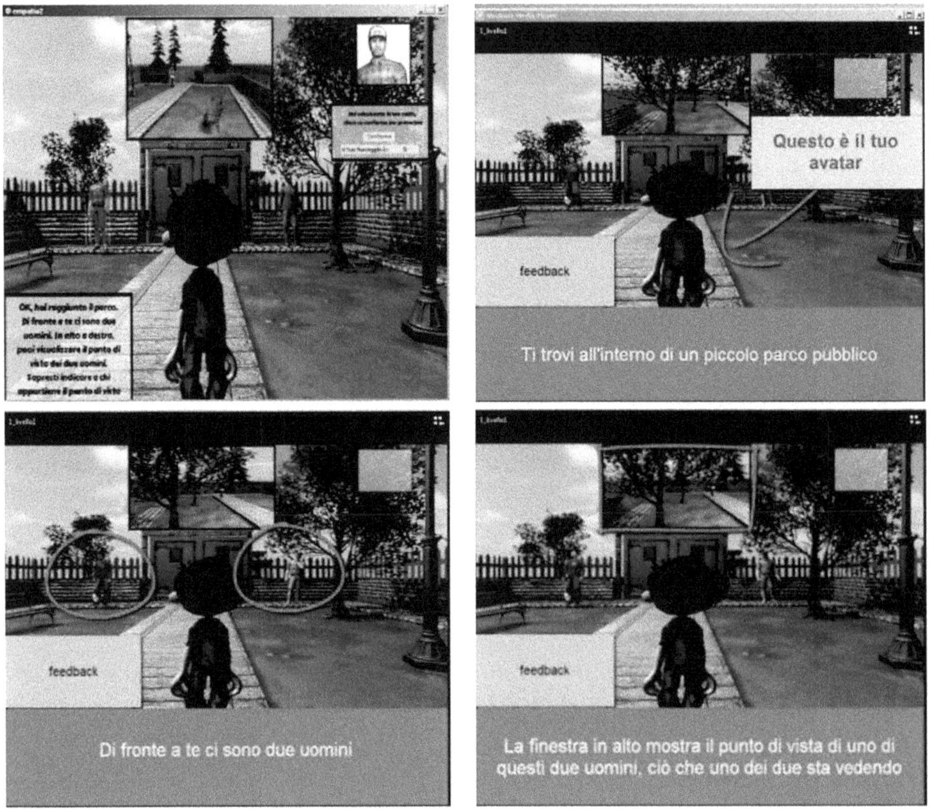

the position of the two men in front of the avatar will be randomly assigned to the two person-placeholders and the location of objects that represent possible landmarks (tree, lamppost, bench) will be randomly assigned to the object-placeholders (Figure 2 - Position of the elements in the game space).

In the second task the user has in front of him one man (Figure 4 - Screenshot of the second task). Two windows in overlay at the top of the screen show two points of view, one of which belongs to the individual in the scene. The user, in this case, must select the window that shows the point of view of the individual present in the park.

In the third task (Figure 5 - Screenshot of the third task) the user is dealing with "the invisible man". The player cannot see the man in the park but he can see, in the overlay window, what the invisible man is seeing. The park is

Figure 2. Position of the elements in the game space

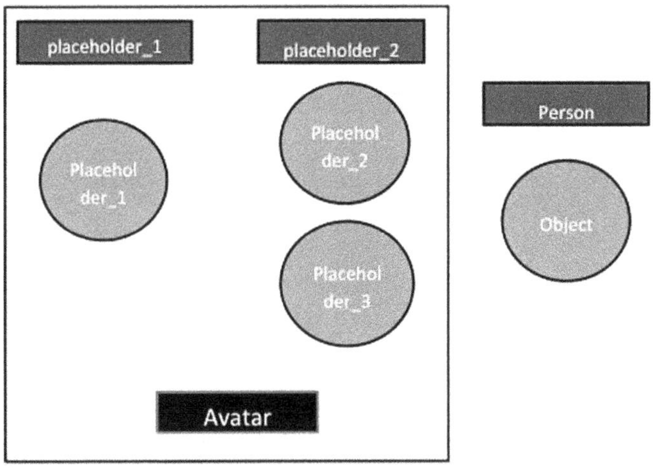

Figure 3. "Semi-subjective point of view," "subjective point of view," "objective point of view"

divided into six zones. By moving the mouse, the player can select the area of the park in which he believes, based on what he sees in the window, he can find the invisible man.

RESULTS

The game was tested between January and March 2015, with a group of 70 children, 35 males and 35 females, from the third, fourth and fifth primary school class and from a first secondary school class of the Istituto Comprensivo

Figure 4. Screenshot of the second task

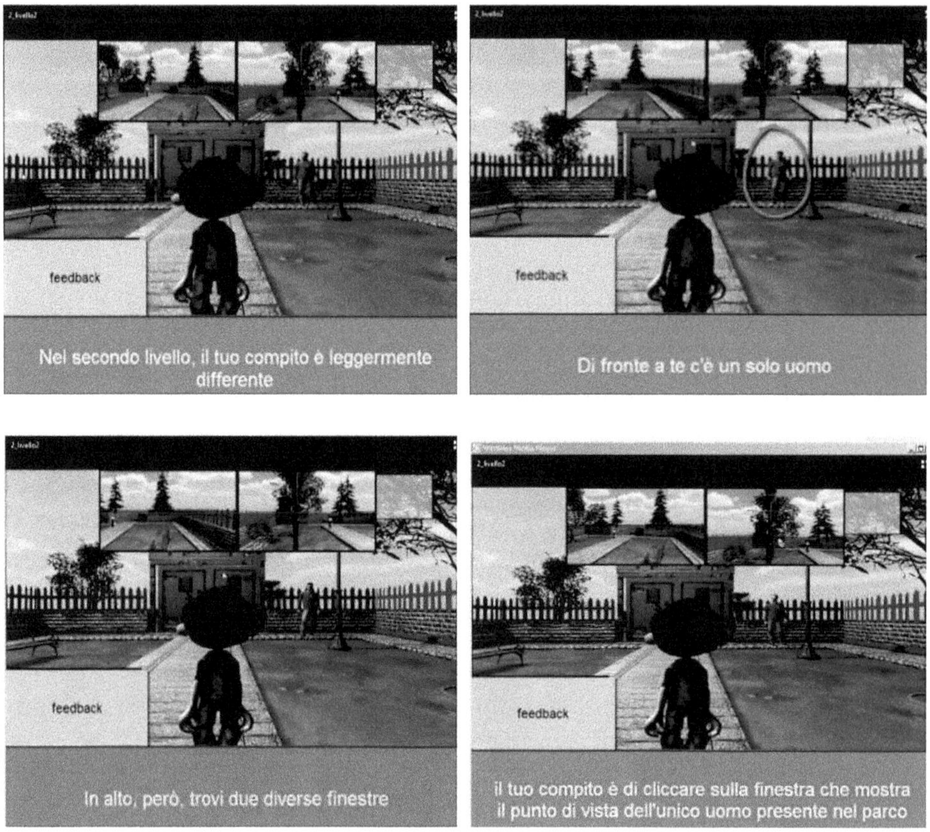

San Valentino Torio, in the province of Salerno. Each user performs 10 attempts for each task. The software records the beginning of each game, the user data (age and gender), and, during the game, time for each attempt and the result (success / failure) of the attempt. Group composition by age and sex, and the overall scores obtained for individual tasks are readable in the Appendix.

The Relationship Between Age and Performance

Analysis of Variance

ANOVA was conducted on the number of successes in task 1 using as BETWEEN FACTOR age and dividing the sample into three groups (6-7 years, 8-9 years, 10-11 years). The results indicate a significant difference in

Figure 5. Screenshot of the third task

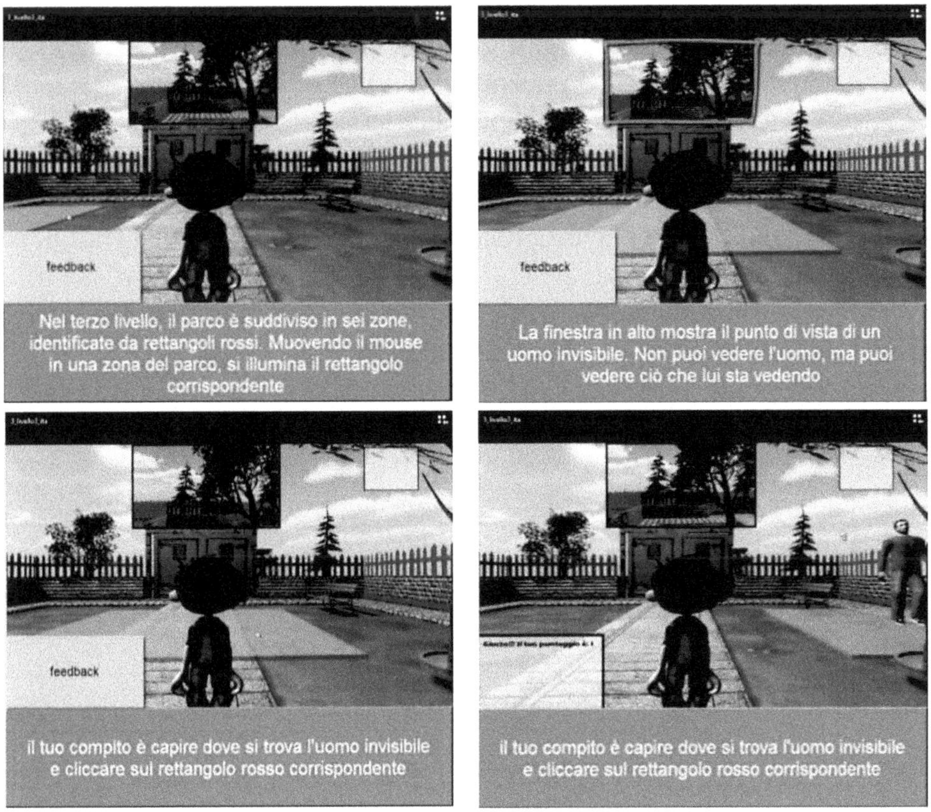

the performance of the subjects relating to the factor taken into consideration [F(2,67)= 60.67, p<0,001] (Table 1).

The data seem to suggest a significant difference in the number of successes achieved in the first task in relation to the age of the subjects.

ANOVA was conducted on the number of successes in task 2 using as BETWEEN FACTOR age and dividing the sample into three groups (6-7 years, 8-9 years, 10-11 years), the results indicate a difference significant in

Table 1. Task 1

Group (Age in Years)	6-7	8-9	10-11
Mean (Score)	5,38	7,52	8,71
SD	1,013	0,911	1,105

the performance of the subjects relating to the factor examined [F(2,67)= 35.84, p<0,001] (Table 2).

The data seem to suggest a significant difference in the number of successes achieved in the second task in relation to the age of the subjects.

ANOVA was conducted on the number of successes in the task 3 using as BETWEEN FACTOR age and dividing the sample into three groups (6-7 years, 8-9 years, 10-11 years), the results indicate a significant difference in the performance of the subjects relating to the factor examined [F(2,67)= 18,43, p<0,001] (Table 3).

The data seem to suggest a significant difference in the number of successes achieved in the third task in relation to the age of the subjects.

Correlation Between Age and Score

Percentages of success in the three tasks were then calculated in relation to the age groups (Table 4).

Table 2. Task 2

Group (Age in Years)	6-7	8-9	10-11
Mean (Score)	4,88	7,0	8,29
SD	1,15	1,39	1,31

Table 3. Task 3

Group (Age in Years)	6-7	8-9	10-11
Mean (Score)	3,75	5,14	6,00
SD	0,90	1,22	1,54

Table 4. % score (10 attempts)

	Age 6-7	**Age 8-9**	**Age 10-11**
Task 1	54%	75%	87%
Task 2	49%	70%	83%
Task 3	38%	51%	60%

Figure 6.

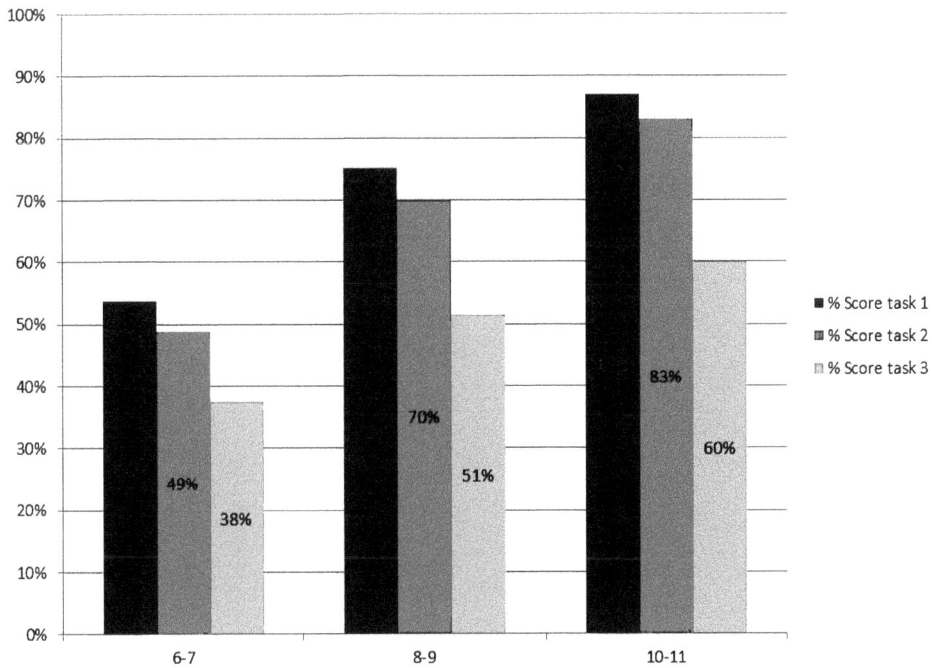

We calculated the Pearson correlation coefficients and coefficients of determination between the variables age (calculated in terms of months) and scores at the task 1 (R= 0.77, R^2=0.58) (Figure 7), age and scores at the task 2 (R= 0.73, R^2=0.54) (Figure 8) age and scores at the task 3 (R= 0.63, R^2=0.39) (Graph 4). The data seem to indicate a strong correlation between age and the scores obtained in the first two tasks and appear to indicate the presence of a moderate correlation with respect to age and the scores obtained in the third task.

The analysis of the data seems to support the hypothesis that between 6 and 11 years the scores to the three tasks significantly increase.

RELATIONSHIP BETWEEN GENDER DIFFERENCE AND PERFORMANCE

In order to assess the impact of gender on the scores we conducted the following hypothesis tests.

Figure 7.

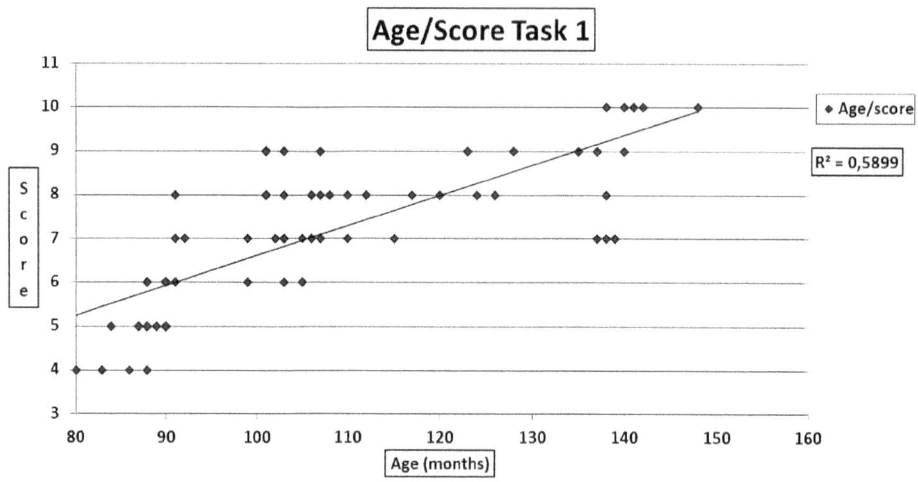

Figure 8.

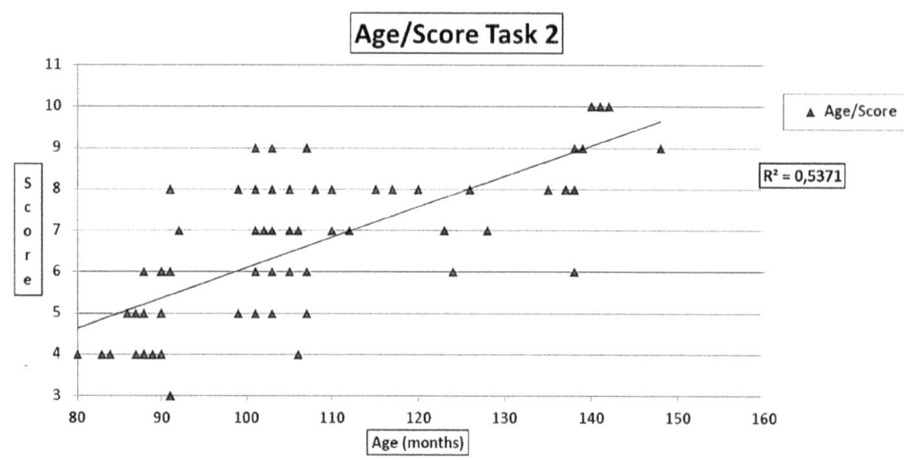

We conducted a T-Student test on scores obtained on at the task 1 by females and males. The results indicate the presence of a significant difference in the performance of the subjects relating to the factor examined (t [68]= 0.961, p = 0.048).

We conducted a T-Student test on scores obtained on at the task 2 by females and males. The results indicate the presence of a significant difference in the performance of the subjects relating to the factor examined (t [68]= 0.964, p = 0.044).

Figure 9.

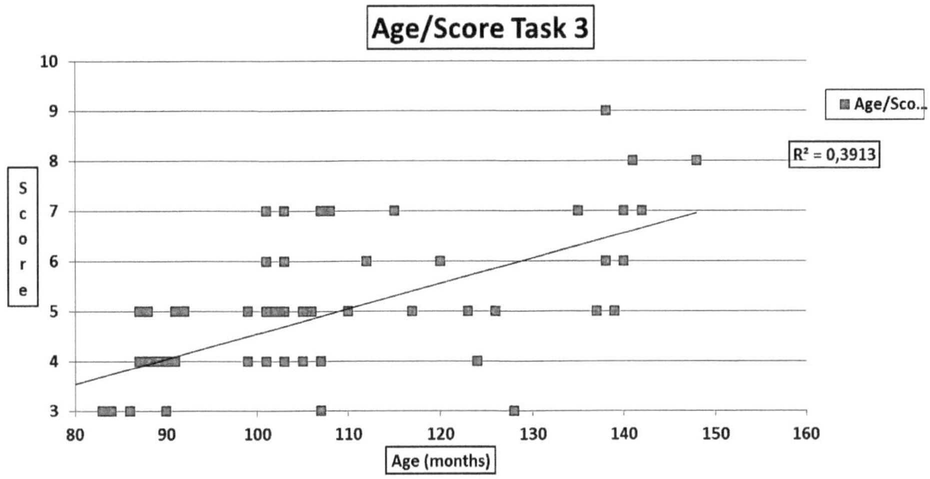

We conducted a T-Student test on scores obtained on at the task 2 by females and males. The results did not highlight significant differences in the performance of the subjects relating to the factor examined (t [68]= 0.705, p = 0.379).

The data seem to indicate the presence of a significant difference in the scores of the first two tasks ($p < 0.05$) in relation to sex, while there were no differences in scores obtained on the third tasks ($p > 0.05$) in relation to sex.

We then calculated the percentage of success in the three tasks in relation to sex and age (Figure 5 and Table 4, Figure 6 and Table 5) and the statistics for the percentage of success in the three tasks in relation to sex and age (Table 6, Table 7).

DISCUSSION

The tests conducted regarding the relationship between age and performance show a significant difference in tasks obtained in relation to the age group selected; the success rates for the three age groups analyzed seem to indicate a gradual improvement in performance in the three tasks in relation to the increase of age.

In particular, the task 3 appears to be the one that presents a greater challenge.

Figure 10.

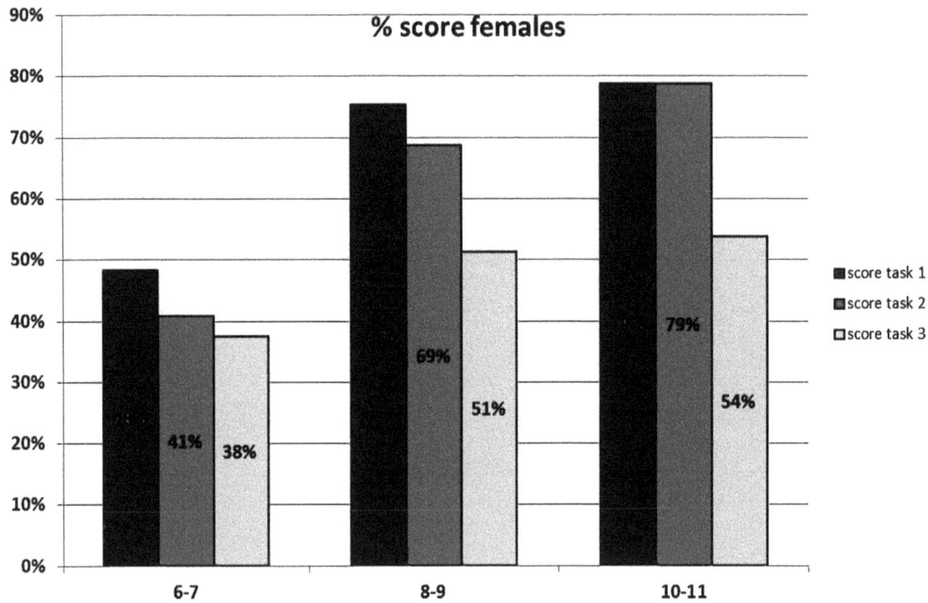

Figure 11.

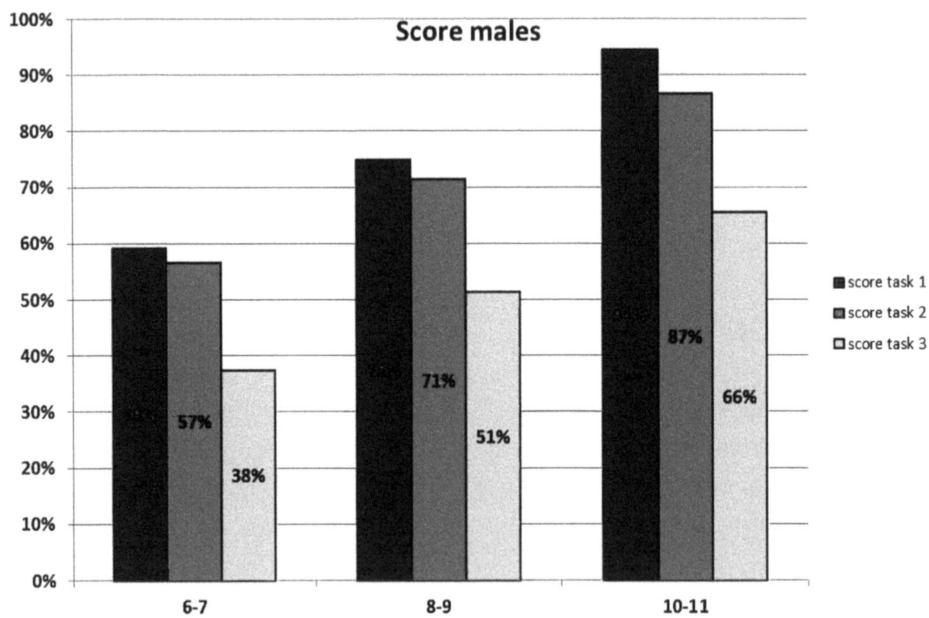

Table 5. % score (females)

	Age 6-7	Age 8-9	Age 10-11
Task 1	48%	75%	79%
Task 2	41%	69%	79%
Task 3	38%	51%	54%

Table 6. % score (males)

	6-7	8-9	10-11
Task 1	59%	75%	94%
Task 2	57%	71%	87%
Task 3	38%	51%	66%

Table 7. Statistics (females)

Age (Years)		6-7	8-9	10-11
Score task 1	Mean	4,8	7,5	7,9
	SD	0,6	0,9	0,8
Score task 2	Mean	4,1	6,9	7,9
	SD	0,5	1,7	1,4
Score task 3	Mean	3,8	5,1	5,4
	SD	0,9	1,3	0,9

Table 8. Statistics (males)

Age (Years)		6-7	8-9	10-11
Score task 1	Mean	5,9	7,5	9,4
	SD	1,1	0,9	0,7
Score task 2	Mean	5,7	7,1	8,7
	SD	1,1	1,0	1,2
Score task 3	Mean	3,8	5,1	6,6
	SD	1,0	1,0	1,8

The tests conducted regarding the relationship between gender and age show a significant difference obtained in the first two tasks in relation to sex. Success rates for the three tasks and sex seem to indicate a gradual improvement in performance in the three tasks in relation to the increase in age for both sexes.

REFERENCES

Berthoz, A. (2011b). *La semplessità*. Torino: Codice.

Hughes, M., & Donaldson, M. (1979). The use of hiding games for studying the coordination of viewpoints. *Educational Review*, *31*(2), 133–140. doi:10.1080/0013191790310207

Lambrey, S., & Berthoz, A. (2007). Gender differences in the use of external landmarks versus spatial representations updated by self-motion. *Journal of Integrative Neuroscience*, *6*(03), 379–401. doi:10.1142/S021963520700157X PMID:17933018

Mitry, J., & King, C. (1997). *The Aesthetics and Psychology of the Cinema*. Indiana University Press.

Piaget, J., & Inhelder, B. (1948). *La représentation de l'espace chez l'enfant*. Paris: Presses Universitaires De France.

Rochat, P. (1995). Perceived reachability for self and for others by 3-to 5-year-old children and adults. *Journal of Experimental Child Psychology*, *59*(2), 317–333. doi:10.1006/jecp.1995.1014 PMID:7722438

Chapter 6
Vicariance, Complexity, Space:
Orienteering for Inclusion

ABSTRACT

The peculiar topic of this research - perspective taking and spatial navigation - has naturally led to cross over the course of the journey other spatial navigation comprehensions. Specifically, during prototype testing, there were several comparisons with another game, Catching Features[1], an orienteering game that tackles the theme of space navigation from a different point of view. The point of view is that of Orienteering, a sports orientation course that consists in making a predefined path with the exclusive help of a compass and a topographic map that represents the details of the territory to be covered. The experimental path of this research has shown that cognitive processes involved in peculiar orienteering activities (map reading, spatial thinking) are also involved in the management of intersubjective relations, and make orienteering a sporting practice with considerable educational potential. This chapter develops by introducing the concept of vicariance, quickly describing orienteering as a sport activity, emphasizing links with Italian national guidelines for the first cycle of education, and, basing on revision of the scientific literature on cognitive processes involved in space navigation and the management of spatial reference systems, provides a proposal for the spread of orientation as an effective didactic practice in a training context aimed at inclusion of pupils Present Special Educational Needs. Paragraph 6.6 discusses the link between autism, empathy and perception – action process, according to neuroscientific evidences.

DOI: 10.4018/978-1-5225-2455-7.ch006

VICARIANCE AND SPATIAL NAVIGATION STRATEGIES

The avoidance that often the so-called normal manifest against those who are disabled is an expression of broadmindedness deficit (Murphy, 2001).

In this quote from The Body Silent, Murphy puts in explicit relationship with disability "the breadth of vision", with the degree of stiffness of the perspective of the actors in a social context. In this work we will support just like orienteering sports involve directly and principally cognitive processes that are critical in the formation of broad-mindedness, ie who are cardinals in acquiring the ability to take the perspective of others.

In practice, the skills involved in the development of cognitive strategies for spatial navigation are skills that allow us to see the world from different points of view, abandoning the egocentric perspective.

The simultaneous manipulation of more spatial reference systems is an essential form of *vicariance*, through which the body solves the local complexity, formulated in terms of the problems posed to the living being on each relevant analysis level (physical-chemical, molecular, synaptic, cognitive, behavioral), through the introduction of an accessory complexity, reduced and re-coded according to the action (Petit, 2012; Sibilio, 2013).

The theoretical framework of this perspective consists of the:

Reversal of the classical description of the mechanisms of perception and action [which] places the intentional and goal-oriented subject at the origin of the process. The subject builds his world according to his basic needs and action tools. This view has also been promoted by Bergson and Husserl. (Berthoz, 2008a)

In this framework, the chapter develops quickly describing orienteering as a sport, and presenting a review of the scientific literature on the cognitive processes involved in spatial navigation and management of spatial reference systems, in the light of the contribution made to this specific field by neuroscience studies. In particular, the work focuses on the classification of cognitive strategies of space navigation, operated by Alain Berthoz in La vicariance (Berthoz, 2013), in four main types: egocentric strategy, allocentric strategy, heterocentric strategy and the maquette3d strategy.

The chapter then concludes with a proposal for the spread of orienteering as an effective teaching practice in an educational context geared to the inclusion of pupils presenting Special Educational Needs.

ORIENTEERING

Golden, Levy and Vohra (1987) provide a detailed definition of the procedural activities to which it is called the athlete who takes part in an orienteering event:

Orienteering is an outdoor sport that is usually played in heavily forested areas. Located in the forest are a number of "control points" each with an associated score. Competitors armed with compass and map are required to visit a subset of the control points from the start point (node 1) so as to maximize their total score and return to the end point (node n) within a prescribed amount of time. (Golden, Levy, & Vohra, 1987)

Given the type of activity, it is easy to imagine how a scientific approach to orienteering involve a wide variety of fields of knowledge (cartography, geography, medicine, sports sciences, education, psychology, neuroscience), which are sometimes considered far apart.

From an educational standpoint, this sport has established since the beginning a tradition of exchange and interaction with school activities:

The educational value provided by orienteering's blend of navigational and physical skills has given it a permanent place in the primary and secondary school curriculum in the United Kingdom. (McNeill, Cory-Wright, & Renfrew, 1998)

In Italy, orienteering is officially recognized by the Ministry of Education with various ministerial documents, including the circular dated 12 November 1998 (ref. 4015 / A1) indicating the terms of cooperation between the Physical Education and Sport Inspectorate and the Italian Federation of Orientation Sports.

This relationship is not surprising, considering that the National Guidelines for Childhood and the First Cycle of Education, among the learning objectives that students should achieve by the third year of secondary school level, cite the ability to orient in the natural environment through map reading and decoding.

Path selection, in fact, is clearly one of the orienteering success factors:

There is no doubt that good route selection and proper execution of the five techniques along your route are keys to success in orienteering. (Ferguson & Turbyfill, 2013)

To be involved in the orientation sports are not only cognitive processes related to space navigation, but also combined task (as well as subtasks) of map reading and navigation (Lobben, 2004).

These tasks are not discrete and separate moments, but are operations that are managed simultaneously by the athlete's brain.

The contribution of neuroscience in the investigation of this specific field of study is filling a gap due to the hybrid nature of the search field, gap that had already been identified and reported years ago.

Just a few years ago, Amy Lobben emphasized how:

Research into cognitive processes in map reading has been conducted primarily in the fields of psychology and cartography.... Cognitive studies and spatial ability measures have been conducted for more than 100 years by psychologists and for more than 30 years by cartographers. However, while some studies provide insight into the cognitive processes and strategies associated with specific map-reading tasks, many of these tasks, strategies, and processes have yet to be identified and, possibly more importantly, understood. (Lobben, 2004)

ORIENTEERING, COMPLEXITY, VICARIANCE: A BRAIN-CENTRIC PERSPECTIVE

Berthoz emphasizes as a condition for use of the multiplicity of spatial references is a coherent perception of body in relation to the environment:

The necessary condition to use the repertoire of spatial reference systems, choosing one, or to address a problem simultaneously from different points of view, is to have a coherent, unique, stable sense of our body and of its relationship with the environment. (Berthoz, 2013)

The concept of space, in this context, must be understood by abandoning Euclidean perspective and by adopting the idea of space as a potential of action:

The space is divided into zones corresponding to actions that can be done. (Berthoz, 2013)

Lewin explains how the approach to the complexity can be linked not only to the description, to the analysis, but also and especially to decision and action:

The molecules in the room I'm now sitting in as I write this chapter are maximally chaotic, and to describe them would require the documentation of the position and activity of every one of them. No simpler description is possible.... That kind of complexity did not interest me. (Lewin, 1999)

Perception is function of action, and perception of a form or of an object is never passive, it is always decision:

When I look at the objects and shapes here, whether it's the room, your faces, your bodies or the objects, my brain does not just analyze them, it decides, it anticipates. (Berthoz, 2011a)[2]

If our brain, in order to decide how to act in a certain situation, were forced to process all the information that the perception of the outside world gives him, the task would be too costly and complex to allow a timely decision: we would be constantly late and we would not be able in the end to produce effective decisions (Rivoltella, 2014).

Faced with insurmountable difficulties posed by a multidimensional reality, the typical activity of living beings proceeds by reducing the dimensionality of the problems, a reduction that addresses the complexity simply by anticipating. The activity of organisms is therefore characterized by a significant reduction of any preconceived complexity and by the introduction of ancillary complexity concerning the local situation.

Berthoz refers to this reduction of complexity as *simplexity*. According to Berthoz:

The brain solves the complexity of the outside world by producing perceptions consistent with the intentions regarding the future, memories of the past and the laws of the outside world that has internalized. (Berthoz, 2012)

Regarding the spatial orientation, the brain can select a reference system based on the task to perform, and during a movement can change the reference.

The multiplicity of the reference systems allows to choose the most suitable system. In the retina, for example, the space is encoded by "retinotopiche" coordinates, in the parietal cortex spatial encoding is egocentric: its reference is the body of the observer. Hippocampus spatial encoding is allocentric, as in a map (Hartley, Lever, Burgess, & O'Keefe, 2014).

We ignore, for the most part, how the brain coordinates all these spaces and uses it to change point of view.

The hypothesis of Berthoz is that the manipulation of spatial referents is an essential form of Vicariance: depending on the context and the task, the brain decides to take as a reference the feet to the ground, or the head if the ground is uneven, or even just simple laying of a finger on a stable element of the environment. Apparently, the management of the different contacts is a further complexity, a detour, but it really simplifies the process of interaction with the environment.

COGNITIVE SPATIAL NAVIGATION STRATEGIES

The vicarious ability to simultaneously use different referents allow us to solve a problem by using different combinations of the reference systems and their neuronal networks, through the use of "cognitive strategies."

Berthoz identifies four different cognitive strategies. Adopting the first strategy, called "egocentric strategy of path", the brain re-build the world around them through "views" subsequent (meetings, events, items noted during the walk), points of view organized in a narrative sequence. The point of view from which develops the analysis of the surrounding reality is in "first person" (in subjective, to use a film lexicon), and therefore it is a fundamentally egocentric process.

The second strategy is an allocentric strategy, based on a mental map of the environment, on which we can follow a route like on a real map. This strategy is allocentric, because it does not use the body of the observer as a reference, but connects together the various elements of the environment in a consistent representation. This is a useful strategy especially about medium-long distances, or for the selection of a path between multiple alternatives. This strategy is the basis of human ability to understand (and developing) the geometry. It should be noted, here, that the appearance of this strategy in the child is related to the appearance of structures such as the hippocampus and the prefrontal cortex.

The third strategy is called heterocentric, and is to describe a path from the point of view of another person (classic example is the request of road traffic information). Conceptually, it is a very similar process to what happens when, during an altercation, we try to understand the point of view of litigants.

The last strategy, useful in the need to locate the position of a portion of space inside an articulated structure (such as an office inside of a building or of a complex interconnected stable), involves the construction of a mental model an three-dimensional structure: the brain processes a representation of the building using a "clear pattern" of the building. Berthoz defines this strategy as "Maquette 3D" (Berthoz, 2013).

TOWARD A SPATIAL THEORY OF EMPATHY

Berthoz proposed a spatial empathy theory, based on the ability of human action on the management point of view. According to this theory:

Empathy is essential to rational thinking, because it allows to examine the facts and arguments from different points of view. (Berthoz, 2011b)

The ability to assume an allocentric perspective is subject to the possibility of performing a:

Mental rotation on ourselves, in relation to the environment, or to an object environment, maintaining a main perspective environment in question. (Berthoz, 2011b)

In practice, it is to be both ourselves and others. Just this is the basic feature of empathy. Empathy is a dynamic process, which requires a doubling. It is, in short, to adopt a point of view egocentric but after an allocentric manipulation, while inhibiting the emotional contagion (which is rather typical of sympathy).

By adopting this approach, empathy is a mode that allows us to relate to each other without attributing to ourselves what the other is experiencing, rather entering the bodies of others to assume the point of view, or to see the world through others' eyes. German philosophy had already emphasized, at the end of the 19th century, this aspect, using the term "Einfühlung", to "feel (fühlen) into (ein)" (Husserl, 1973; Vischer, 1927).

In Einfühlung, the prefix *ein* refers to a space process of mental simulation through which we imagine to be placed in the body of others (spatial aspect).

The root *fühlen* instead refers to a subjective experience of sensory or emotional nature. The same etymological root, therefore, suggests a dynamic interaction between the simultaneous experiences of our own bodies and the bodies of others (Berthoz & Thirioux, 2010).

In fact, this approach represents a reversal with respect to neuroscientific perspective that "addressed the question of the neural basis of sympathy and emotion via emotional contagion and resonance and do not address the complex dynamic mechanisms of empathy" (Berthoz & Thirioux, 2010).

In the hypothesis developed at the College de France, four processes were identified at the base of the empathic relationships:

1. The construction of a coherent sense of our body and of its relationship with the environment.
2. The capacity to resonate with the emotions and perceptions of others.
3. The ability to change the point of view or perspective and to move our bodies and our brains in the body and brain of others ("Einfühlung").
4. The capacity to abandon the egocentric or heterocentric perspective (our point of view or the point of view of others) to adopt an allocentric perspective, inhibiting the emotional contagion (Berthoz & Thirioux, 2010).

The empathic processes require the contribution (albeit not exclusively) of different brain mechanisms involved in spatial perception, in the mental manipulation of the reference systems and in the change of perspective (Berthoz & Thirioux, 2010). At the basis of empathy, in this case, we find the kinesthesia, because it contributes to intersubjectivity, as understanding, for analog transfer in the body of another agent, of a part of the intentional actions of the subject and of the intentions that precede and accompany those actions

In essence, as in space navigation we are useful to change spatial referents and even the reference systems, so in social life often we change referent by taking others as a reference.

Autism Spectrum as a Spatial Empathy Disorder?

The idea of autism spectrum as an empathy disorder extends the most dated ToM (Theory of Mind) / mind-blindness hypothesis. Theory of mind is the ability to attribute mental states to others, to infer what someone else is thinking or feeling, roughly attributable to the component that was generally defined 'cognitive empathy', distinct from the 'affective empathy', i.e., the ability to respond with an appropriate emotion to the mental states of someone else.

It is through the bodily experience and physical activity that the individual first senses the presence of the other and then comprises the actions of others. In other words, the kinesthesia is the basis of empathic processes, as understanding, for analog transfer in the body of another agent, of a part of intentional actions performed by the subject and the intentions that precede and accompany those actions.

In this hypothesis, the fundamental characteristic of empathy consists in being both ourselves and the other. Adopting this perspective, the ability to assess the perceptual egocentrism condition is of fundamental importance in the teaching - learning process.

Empathy implies acknowledgment and understanding of one another's emotional state. According to Jean Decety, is the organization itself of our nervous system to provide the basic mechanism for resonance with others, the ability to simulate our actions and their consequences, as well as other people's actions and their consequences (Decety & Jackson, 2004).

This mechanism proposed by Decety and Jackson (2004), of the shared representation (neural patterns temporarily activated from the real perception or evoked from memory), driven by the common code between perception and action, provides the default mode for relating to others implicitly. This mechanism is based on the idea that the perception of a certain behavior in another individual automatically activates our representations of such behavior. The mechanism involves the fundamental physiological properties of the nervous system with regard to the continuity between the action and cognition.

In particular, Decety and Jackson move away from meaning of empathy gained in some areas of psychology, who stressed the inductive, sometimes almost mechanical, aspect of empathy, to mark the intentional aspect of empathic process:

Comparative psychologists view empathy as a kind of induction process by which emotions, both positive and negative, are shared and by which the probabilities of similar behavior are increased in the participants. In our view, this is not a sufficient mechanism to account for human empathy. Feelings may be shared, but humans are able to intentionally "feel for" and act on behalf of other people whose experiences differ greatly from their own. (Decety & Jackson, 2004)

The theoretical framework whereby neurosciences, and in particular Decety and Jackson, move makes direct reference to a paradigm of perception / action that now appears permanently reversed with respect to the idea of perception - action that was dominant in the second half of the twentieth century, solidly hinged on cognitivist foundations and, in particular, on the computational metaphor.

Decety and Jackson explicitly link the concept of empathy to the process of perception - action and the concept of shared representation: in neuroscience, evidence for this perception/action coupling ranges from electrophysiological recordings in monkeys, in which mirror neurons that fire both during goal-directed actions and observation of actions performed by another individual (Rizzolatti, Fogassi, & Gallese, 2001), to functional neuroimaging experiments in humans, which demonstrate that the neural circuit involved in action execution overlaps with that activated when actions are observed (Decety & Jackson, 2004).

This hypothesis is supported by abundant scientific literature in the field of perception and action, entitled under the name of common coding theory (Prinz, 1997). The basic assumption is that actions are coded in terms of the perceptible effects that are expected to generate. These representations, then, can be shared among different individuals. In other words, the isomorphism between representation and action allows the individual to implicitly learn the goal of others through the use of action representation system. Through bodily experience and physical activity, therefore, the individual perceives the presence of the other and then comprises the actions of others (Gallagher, 2001).

One of the main components of empathy is based on the mental simulation of another's subjectivity that occurs, at least in humans, under conscious control. In fact, the activation of the shared representation mechanism, without self-awareness and emotional regulation, produces no empathy, but anxiety or distress.

In humans, the empathy construct is a more complex psychological state than that associated with the automatic sharing of emotions. As in other species, emotions and feelings can be shared between individuals, but humans are also able to "feel" and act intentionally on behalf of other persons whose experiences can differ greatly from them (Decety, 2009).

Empathy, in this sense, is to be considered an intentional skill.

According to the DSM-V, the diagnosis of "autism spectrum disorder" requires the presence of at least three symptoms in the category of "deficit of social communication" and at least two in that of "repetitive behaviors."

Children with autism, therefore, show a large deficit area of social communication, and most scholars agree that the lack of empathy figures prominently among these disorders (Frith, 2001).

The cause of this deficit is, however, controversial. Decety & Jackson offer a complete and concise overview of the different positions (Decety & Jackson, 2004).

Baron-Cohen, Leslie, and Frith (1985) have argued that the social difficulties in autism arise from a problem in a mentalizing mechanism; their work, in fact, is mainly based on the construct of the theory of mind. "Theory of Mind" in the scientific literature relata autism, refers to "the ability of [the neurotypical] to attribute mental states (such as beliefs, desires, intentions, etc.) to themselves and other people, as a way of making sense of and predicting behaviour" (Baron-Cohen, Tager-Flusberg, & Cohen, 1994). They argue that the severe forms of autism can be characterized by mindblindness - inability to accurately infer, or perhaps even to recognize the existence of thoughts and feelings in other people (Ickes, 2009).

Other authors believe that children with autism have difficulty feel and express emotions and that this basic deficit prevent them from engaging in social interactions (Hobson, 1989). Yet others argue that deficits in executive functions are the main cause for the observed social disorders in autism (Russell, 2013). Rogers e Pennington (Rogers & Pennington, 1991) have suggested a waterfall model in which the lack of some aspects of interpersonal development in a prior stage upsets some developments in the later stages. Finally, Dawson (1991) proposed that autism involves impaired attention to social cues such as facial expressions, language, gestures (Dawson, 1991).

Decety and Jackson conclude that seems likely that children with autism exhibit difficulties in taking the perspective of others - activities that requires executive resources - but they have the required physiological substrate to show of emotion sharing capabilities. In addition, several studies prove some deficit in the perception - action process (Charman et al., 1997). This deficit in the motor resonance can be the central node of a profound impairment of managing emotions. Yet, several studies suggest that a poor mirror system (MNS mirron Neuron System) may contribute to the motor and social problems experienced by people with autism spectrum disorders (Fadiga, Fogassi, Pavesi, & Rizzolatti, 1995). The activation to a lesser extent of the mirror system in the motor cortex can be in part responsible for the social cognition deficit, in particular as it regards self-other representation, social skills, and the development of empathy (Dapretto et al., 2006).

These latest results seem to suggest that malfunctions of the mirror system hinder the normal development of self - other relationship, triggering waterfall processes that result in empathy deficit. What is, however, the dysfunctional element of empathy in autism is still unclear, and seems to be no sufficient empirical studies that have addressed the subject. According to available research, seem to be affected both fundamental mechanisms of empathy: the aspect linked to the perception process - action process and the metacognitive aspect of self-regulation of emotions. In any case, the literature is not sufficient to establish the extent to which these elements are altered in people with autism.

ORIENTEERING AND SPECIAL EDUCATIONAL NEEDS

According to Fabio Dovigo:

In Italy, in recent years, the term "inclusion" has gradually begun to replace, in the documents and formal and informal speeches, the traditional one of "integration".... The two expressions refer to two very different educational scenarios. (Dovigo, 2008).

The paradigm that implicitly includes the idea of integration is an "assimilationist" paradigm, based on the adaptation of pupils with disabilities to school organization that is fundamentally structured according to 'normal' students, where the design for special "students" still plays a marginal or residual role.

R. F. Murphy directly relates disability with the "breadth of vision", with the degree of stiffness of the perspective of the actors in a social context (Murphy, 2001).

In the scientific literature, several times orienteering has been listed as a learning resource with a recognized potential for integration and inclusion.

The reasons for this "inclusive potential" were gradually detected in not peculiar orienteering features (outdoor activities, competition), in general or in relation to particular types of disabilities (above all, orienteering as a tool for the inclusion of blind child).

The sport of orienteering, however, goes beyond these features. Orienteering directly and principally involves cognitive processes that are crucial in the acquisition of the ability to take the perspective of others.

Cognitive processes involved in specific activities of orienteering (map reading, identification of pathways, spatial thinking) are processes involved in the management of intersubjective relationship. In essence, the skills involved in the reading of maps and in the development of space navigation strategies are skills that allow us to see the world from different points of view, leaving the egocentric perspective, and are therefore involved in inclusion-oriented educational strategies.

Going out from egocentrism means being able to see the reality (not only spatial reality, but also a mental reality, an linguistic reality or cultural) from several points of view, however, other than their own. (Trisciuzzi & Zappaterra, 2011)

Orienteering, therefore, takes the form of effective teaching practice in an educational context directed to the inclusion of pupils presenting Special Educational Needs.

REFERENCES

Baron-Cohen, S., Leslie, A. M., & Frith, U. (1985). Does the autistic child have a theory of mind? *Cognition*, *21*(1), 37–46. doi:10.1016/0010-0277(85)90022-8 PMID:2934210

Baron-Cohen, S. E., Tager-Flusberg, H. E., & Cohen, D. J. (1994). *Understanding other minds: Perspectives from autism.* Paper presented at the Workshop in Seattle, Seattle, WA.

Berthoz, A. (2008a). *Neurobiology of "Umwelt": How Living Beings Perceive the World.* Springer.

Berthoz, A. (2011a). Fondements cognitifs de la perception de l'espace. In J.-F. Augoyard (Ed.), Faire une ambiance = creating an atmosphere: actes du colloque international Grenoble 10-12 septembre 2008 (pp. 121-132). Grenoble: A la croisée.

Berthoz, A. (2011b). *La semplessità.* Torino: Codice.

Berthoz, A. (2012). *Simplexity: Simplifying Principles for a Complex World* (G. Weiss, Trans.). Yale University Press. doi:10.12987/yale/9780300169348.001.0001

Berthoz, A. (2013). *Vicariance (La): Le cerveau créateur de mondes*. Odile Jacob.

Berthoz, A., & Thirioux, B. (2010). A Spatial and Perspective Change Theory of the Difference Between Sympathy and Empathy. *Paragrana, 19*(1), 32–61. doi:10.1524/para.2010.0003

Charman, T., Swettenham, J., Baron-Cohen, S., Cox, A., Baird, G., & Drew, A. (1997). Infants with autism: An investigation of empathy, pretend play, joint attention, and imitation. *Developmental Psychology, 33*(5), 781–789. doi:10.1037/0012-1649.33.5.781 PMID:9300211

Dapretto, M., Davies, M. S., Pfeifer, J. H., Scott, A. A., Sigman, M., Bookheimer, S. Y., & Iacoboni, M. (2006). Understanding emotions in others: Mirror neuron dysfunction in children with autism spectrum disorders. *Nature Neuroscience, 9*(1), 28–30. doi:10.1038/nn1611 PMID:16327784

Dawson, G. (1991). VIII A Psychobiological Perspective on the Early Socio-emotional Development of Children with Autism. *Models and Integrations, 3*, 207.

Decety, J. (2009). *The social neuroscience of empathy*. Cambridge, MA: MIT Press. doi:10.7551/mitpress/9780262012973.001.0001

Decety, J., & Jackson, P. L. (2004). The functional architecture of human empathy. *Behavioral and Cognitive Neuroscience Reviews, 3*(2), 71–100. doi:10.1177/1534582304267187 PMID:15537986

Dovigo, F. (2008). L'Index per l'inclusione: una proposta per lo sviluppo inclusivo della scuola. In L'Index per l'inclusione. Trento: Edizioni Erickson.

Fadiga, L., Fogassi, L., Pavesi, G., & Rizzolatti, G. (1995). Motor facilitation during action observation: A magnetic stimulation study. *Journal of Neurophysiology, 73*(6), 2608–2611. PMID:7666169

Ferguson, C., & Turbyfill, R. (2013). *Discovering Orienteering: Skills, Techniques, and Activities*. Human Kinetics.

Gallagher, S. (2001). The practice of mind. Theory, simulation or primary interaction? *Journal of Consciousness Studies, 8*(5-7), 83–108.

Golden, B. L., Levy, L., & Vohra, R. (1987). The orienteering problem. *Naval Research Logistics, 34*(3), 307-318.

Hartley, T., Lever, C., Burgess, N., & OKeefe, J. (2014). Space in the brain: How the hippocampal formation supports spatial cognition. *Philosophical Transactions of the Royal Society of London. Series B, Biological Sciences*, *369*(1635), 20120510. doi:10.1098/rstb.2012.0510 PMID:24366125

Hobson, R. P. (1989). On sharing experiences. *Development and Psychopathology*, *1*(03), 197–203. doi:10.1017/S0954579400000390

Husserl, E. (1973). *Zur Phänomenologie der Intersubjektivität: Texte aus dem Nachlaß. Zweiter* (Vol. 12). Springer. doi:10.1007/978-94-010-2474-7

Ickes, W. (2009). Empathic accuracy: Its links to clinical, cognitive, developmental, social, and physiological psychology. *The Social Neuroscience of Empathy*, 57-70.

Lewin, R. (1999). *Complexity: Life at the edge of chaos*. Chicago: The University of Chicago Press.

Lobben, A. K. (2004). Tasks, Strategies, and Cognitive Processes Associated With Navigational Map Reading: A Review Perspective. *The Professional Geographer*, *56*(2), 270–281.

McNeill, C., Cory-Wright, J., & Renfrew, T. (1998). *Teaching orienteering*. ERIC.

Murphy, R. F. (2001). *The Body Silent*. Norton.

Petit, J. (2012). *Complexité-Simplexité - De la simplexité au champ phénoménal: La réponse du vivant à la complexité*. Retrieved from http://www.jeanluc-petit.com/sites/default/files/seminaire_complexite-simplexite_23-24.05.12.pdf

Prinz, W. (1997). Perception and action planning. *The European Journal of Cognitive Psychology*, *9*(2), 129–154. doi:10.1080/713752551

Rivoltella, P. C. (2014). *La previsione. Neuroscienze, apprendimento, didattica*. La Scuola.

Rizzolatti, G., Fogassi, L., & Gallese, V. (2001). Neurophysiological mechanisms underlying the understanding and imitation of action. *Nature Reviews. Neuroscience*, *2*(9), 661–670. doi:10.1038/35090060 PMID:11533734

Rogers, S. J., & Pennington, B. F. (1991). A theoretical approach to the deficits in infantile autism. *Development and Psychopathology*, *3*(02), 137–162. doi:10.1017/S0954579400000043

Russell, J. (2013). *Agency: Its role in mental development*. Psychology Press.

Sibilio, M. (2013). *La didattica semplessa*. Napoli: Liguori.

Trisciuzzi, L., & Zappaterra, T. (2011). Dislessia, disgrafia e didattica inclusiva. *ANNALI DELLA PUBBLICA ISTRUZIONE*, *2*, 51–76.

Vischer, R. (1927). Über das optische Formgefühl (1872); Der ästhetische Akt und die reine Form (1874): Über ästhetische Naturbetrachtung (1890). *Drei Schriften zum ästhetischen Formproblem–Leipzig*.

ENDNOTES

1 http://www.catchingfeatures.com

2 Quand je regarde les objets et les formes ici, que ce soit la salle, vos visages, vos corps ou les objets, mon cerveau ne se contente pas de les analyser, il décide, il anticipe. (Berthoz, 2011a)

Conclusion

THE BODY AND THE TEXT

This book proposes a reading that overcomes the classic description of the mechanism of action and perception, returning to the acting subject the centrality of this process. Perception, it has been said, is a function of action, there is no perception that does not refer to the body in action. It is a thesis that fits in a more general context. The human body has been the focus of attention over the last few decades that has led it to be the theoretical place for encountering (and clashing) between different sciences and different scientific research trajectories. From philosophy to medicine, from neuroscience to anthropology, from law to pedagogy, the disciplines that place man as the subject of his own investigation have all claimed, from their own peculiar epistemic point of view, the centrality of the body. In this text, the body in action was meant as an interface, as a coding and decoding device. In other words, it has become a text.

Knowledge gained during evolution is stored genetically, expressed in the functional brain architecture and made complete by the learning process. This a priori knowledge determines how we perceive and also what we perceive, defines the sensory categories according to which we attribute quality to our experiences.

These attributions:

Are rather idiosyncratic. Thus, we arbitrarily classify electromagnetic radiation with wavelengths between 400 to 700 nanometers as light, because the photoreceptors in the eye are sensitive to this wavelength range. Radiations with slightly longer wavelengths stimulate our temperature receptors and we categorize the respective sensations as temperature. A similar arbitrariness of category boundaries is observable in other sensory domains. A priori knowledge is also required for the definition of the nature of objects. We have clear concepts of what an object is, and the corresponding rules were worked out in great detail by the Gestalt psychologists at the beginning of the last century. Objects should be confined by continuous borders, should be solid and, if they move, all components should move in a coherent way. (Singer, 2009)

Human beings have evolved in similar environments and therefore share the same cognitive mechanisms. We perceive the world in the same way and tend to agree on most of its attributes. Cultural evolution, however, has created a growing socio-cultural environment that contributed to the epigenetic formation of brain architectures. Thus, epigenetic features have undergone to significant changes (Singer, 2009).

If the perception of social realities is also determined by implicit knowledge, acquired through early imprinting and education, it is plausible that members of different cultures perceive the attributes of the socio-cultural world in radically different ways, without however being aware. This dependence of perception on implicit knowledge is probably at the root of a historical exclusion path that has produced, according to Foucault (Foucault & Ferrucci, 1980), a liminar condition.

Inclusion appears as one of the pertinent emergencies of all disciplines that place the human being as an object of investigation. The body is the pivot of this investigation.

In a work on Posthuman Bodies, Steven Shaviro states:

Postmodernism is distinguished...by a propensity to invent new organs of perception and action, as Burroughs, McLuhan, David Cronenberg, Michael Taussig, and Donna Haraway all in various ways recommend. (Shaviro, 1997)

Except for any consideration on the idea of postmodern, Shaviro identifies a trend, the "propensity to invent new organs of perception / action", documentable along two lines, the technological and biological, which have between them wide spaces of overlapping and contamination. In Donna Haraway's "Cyborg Manifesto" (Haraway, 1985), in fact, the common denominator of both lines is the concept of "code".

Whether it is redefined according to the technological grafts, or is analyzed and modified according to the characteristics of the biological sciences, the body is a text.

Boundaries jump into a perpetual decoding and recoding process (transcoding), whose object (environment) loses every ontological claim to be a readable code in accordance with its biological devices, where code snippets are read differently from Different devices.

There is no "out of there" for everyone, the body builds our "reality" by constantly transcoding fragments of code, producing, as Agamben writes about von Uexkull, "an infinite variety of perceptive worlds...inconsistent and mutually exclusive" (Agamben, 2002).

Space and time are read (and written) codes through our biological structure, there is no space and a unique time, there is no scenario in which all organisms move.

There is a functional unit of the body with some elements of the surrounding environment, there is an interface with the space and time that is calibrated on the biological structure, and it changes when this structure changes. The nature of this interface is of a semantic nature, it fully involves the epistemological plan:

My nervous system does not, indeed, cannot, tell me what is "out there", not because of mechanical but because of logico-semantical reasons. My nervous system cannot tell me anything because it is "me": I am the activity of my nervous system; all my nervous system talks about is its own state of sensory-motor activity. (Von Foerster, 1984)

The attention, necessarily, moves from the text to the body, since it the same text.

From these premises the work presented here has just started, to immediately come to the need to equip itself with his own investigation tools.

These pages are certainly not the report of a finished work, nor are they merely a theoretical reflection. More than anything, this work is a position question is, the opening of a research track.

The attempt made with this volume consists, then, in the reporting of an ongoing research, justifying the rationale and conceptual framework in a broader context than that offered by traditional paper.

Research is only at the beginning, because the topic dealt with is, as it is said, in an ideal crossroads between several disciplines (philosophy, pedagogy, sports sciences, neuroscience, psychology) and the deviations that occur on the path are all worthy of attention.

REFERENCES

Agamben, G. (2002). *L'aperto: l'uomo e l'animale*. Bollati Boringhieri.

Foucault, M., & Ferrucci, F. (1980). *Storia della follia nell'età classica*. Rizzoli.

Haraway, D. J. (1985). *A manifesto for cyborgs: Science, technology, and socialist feminism in the 1980s*. Center for Social Research and Education.

Shaviro, S. (1997). *Doom patrols: a theoretical fiction about postmodernism*. Serpent's Tail.

Singer, W. (2009). The Brain's View of the World Depends on What it has to Know. In A. Berthoz (Ed.), *Neurobiology of "Umwelt"* (pp. 39–52). Berlin: Springer. doi:10.1007/978-3-540-85897-3_5

Von Foerster, H. (1984). *Observing Systems*. Intersystems Publ.

Appendix

GROUP COMPOSITION BY AGE AND SEX: SCORES FOR INDIVIDUAL TASKS

The game was tested between January and March 2015, with a group of 70 children, 35 males and 35 females, from the third, fourth and fifth primary school class and from a first secondary school class of the Istituto Comprensivo San Valentino Torio, in the province of Salerno. Each user performs 10 attempts for each task. The software records the beginning of each game, the user data (age and gender), and, during the game, time for each attempt and the result (success / failure) of the attempt. Table 1 shows group composition by age and sex, and the overall scores obtained for individual tasks.

Table 1. Group composition by age and sex, and the overall scores obtained

Age (Months)	Age (Years)	Sex	Score Task 1	Score Task 2	Score Task 3
80	6,7	f	4	4	2
83	6,9	f	4	4	3
84	7,0	m	5	4	3
86	7,2	m	4	5	3
87	7,3	f	5	4	5
87	7,3	m	5	5	4
88	7,3	f	4	4	5
88	7,3	f	5	4	4
88	7,3	f	5	4	4
88	7,3	m	5	5	4
88	7,3	m	6	6	4
88	7,3	m	6	5	5
89	7,4	f	5	4	4

continued on following page

Table 1. Continued

Age (Months)	Age (Years)	Sex	Score Task 1	Score Task 2	Score Task 3
90	7,5	f	5	5	3
90	7,5	f	5	4	4
90	7,5	f	5	5	3
90	7,5	f	5	4	4
90	7,5	m	6	6	2
90	7,5	m	6	5	3
90	7,5	m	6	6	3
91	7,6	f	6	3	4
91	7,6	m	8	6	4
91	7,6	m	7	8	5
92	7,7	m	7	7	5
99	8,3	f	7	8	4
99	8,3	f	6	5	5
101	8,4	f	9	5	7
101	8,4	f	8	6	5
101	8,4	f	9	8	4
101	8,4	m	8	6	6
101	8,4	m	8	7	7
101	8,4	m	8	9	4
102	8,5	f	7	7	5
103	8,6	f	7	5	4
103	8,6	m	7	6	5
103	8,6	m	9	7	4
103	8,6	m	6	8	6
103	8,6	m	8	9	7
105	8,8	m	6	6	5
105	8,8	m	7	8	5
105	8,8	m	6	7	4
106	8,8	f	7	4	5
106	8,8	m	8	7	5
107	8,9	f	9	9	7
107	8,9	f	7	9	4
107	8,9	f	7	5	3
107	8,9	m	8	6	3
108	9,0	f	8	8	7

continued on following page

Table 1. Continued

Age (Months)	Age (Years)	Sex	Score Task 1	Score Task 2	Score Task 3
110	9,2	f	7	8	5
110	9,2	m	8	7	5
112	9,3	m	8	7	6
115	9,6	f	7	8	7
117	9,8	f	8	8	5
120	10,0	m	8	8	6
123	10,3	m	9	7	5
124	10,3	f	8	6	4
126	10,5	f	8	8	5
128	10,7	m	9	7	3
135	11,3	m	9	8	7
137	11,4	f	9	8	5
137	11,4	f	7	8	5
138	11,5	f	7	6	6
138	11,5	f	8	8	6
138	11,5	m	10	9	9
139	11,6	f	7	9	5
140	11,7	f	9	10	7
140	11,7	m	10	10	6
141	11,8	m	10	10	8
142	11,8	m	10	10	7
148	12,3	m	10	9	8

Related Readings

To continue IGI Global's long-standing tradition of advancing innovation through emerging research, please find below a compiled list of recommended IGI Global book chapters and journal articles in the areas of spatial navigation, cognition, and user interfaces. These related readings will provide additional information and guidance to further enrich your knowledge and assist you with your own research.

Abeykoon, A. H., & Ruwanthika, R. M. (2016). Remote Gripping for Effective Bilateral Teleoperation. In J. Rodrigues, P. Cardoso, J. Monteiro, & M. Figueiredo (Eds.), *Handbook of Research on Human-Computer Interfaces, Developments, and Applications* (pp. 99–134). Hershey, PA: IGI Global. doi:10.4018/978-1-5225-0435-1.ch005

Agar, D. A., & Chappell, P. J. (2016). The Creation of a Rubric for the Evaluation of Language Teaching and Learning Videogames. In D. Russell & J. Laffey (Eds.), *Handbook of Research on Gaming Trends in P-12 Education* (pp. 353–379). Hershey, PA: IGI Global. doi:10.4018/978-1-4666-9629-7.ch017

Ahmed, M. S. (2016). Framework to Develop a Learning Analytics System for Smartphone Blended Learning Environment. In L. Briz-Ponce, J. Juanes-Méndez, & F. García-Peñalvo (Eds.), *Handbook of Research on Mobile Devices and Applications in Higher Education Settings* (pp. 72–91). Hershey, PA: IGI Global. doi:10.4018/978-1-5225-0256-2.ch004

Ahuja, S. (2017). Using the Flipped Classroom to Improve Knowledge Creation of Master's-Level Students in Engineering. In T. Kidd & L. Morris Jr., (Eds.), *Handbook of Research on Instructional Systems and Educational Technology* (pp. 326–339). Hershey, PA: IGI Global. doi:10.4018/978-1-5225-2399-4.ch028

Akcaoglu, M., Gutierrez, A. P., Hodges, C. B., & Sonnleitner, P. (2017). Game Design as a Complex Problem Solving Process. In R. Zheng & M. Gardner (Eds.), *Handbook of Research on Serious Games for Educational Applications* (pp. 217–233). Hershey, PA: IGI Global. doi:10.4018/978-1-5225-0513-6.ch010

AlKassim, Z. S., & Mohamed, N. (2016). Sixth Sense Technology: Exploring Future Opportunities in Human Computer Interaction. In J. Rodrigues, P. Cardoso, J. Monteiro, & M. Figueiredo (Eds.), *Handbook of Research on Human-Computer Interfaces, Developments, and Applications* (pp. 188–215). Hershey, PA: IGI Global. doi:10.4018/978-1-5225-0435-1.ch008

Almajano, P., Lopez-Sanchez, M., Rodriguez, I., Puig, A., Llorente, M. S., & Ribera, M. (2016). Training Infrastructure to Participate in Real Life Institutions: Learning through Virtual Worlds. In F. Neto, R. de Souza, & A. Gomes (Eds.), *Handbook of Research on 3-D Virtual Environments and Hypermedia for Ubiquitous Learning* (pp. 192–219). Hershey, PA: IGI Global. doi:10.4018/978-1-5225-0125-1.ch008

Amadieu, F., Pecoste, C., Mariné, C., van de Leemput, C., & Lescarret, C. (2016). Effects of Studying Tasks Compatibility with Tablets on Their Acceptance: How Experienced Tasks with Tablets Can Modify Perceptions of Tablets. In F. Neto, R. de Souza, & A. Gomes (Eds.), *Handbook of Research on 3-D Virtual Environments and Hypermedia for Ubiquitous Learning* (pp. 338–361). Hershey, PA: IGI Global. doi:10.4018/978-1-5225-0125-1.ch014

Ambler, T., Breyer, Y., & Young, S. (2014). Piloting Online Submission and Online Assessment with GradeMark. In S. Kennedy-Clark, K. Everett, & P. Wheeler (Eds.), *Cases on the Assessment of Scenario and Game-Based Virtual Worlds in Higher Education* (pp. 125–151). Hershey, PA: IGI Global. doi:10.4018/978-1-4666-4470-0.ch004

Andreu, L., & Sanz-Torrent, M. (2017). The Visual World Paradigm in Children with Spoken Language Disorders. In C. Was, F. Sansosti, & B. Morris (Eds.), *Eye-Tracking Technology Applications in Educational Research* (pp. 262–282). Hershey, PA: IGI Global. doi:10.4018/978-1-5225-1005-5.ch013

Ang, R. P., Tan, J. L., Goh, D. H., Huan, V. S., Ooi, Y. P., Boon, J. S., & Fung, D. S. (2017). A Game-Based Approach to Teaching Social Problem-Solving Skills. In R. Zheng & M. Gardner (Eds.), *Handbook of Research on Serious Games for Educational Applications* (pp. 168–195). Hershey, PA: IGI Global. doi:10.4018/978-1-5225-0513-6.ch008

Annan, N. K. (2016). Stump: A Model for the Adoption, Implementation, and Use of M-Learning or E-Learning Platform. In L. Briz-Ponce, J. Juanes-Méndez, & F. García-Peñalvo (Eds.), *Handbook of Research on Mobile Devices and Applications in Higher Education Settings* (pp. 344–372). Hershey, PA: IGI Global. doi:10.4018/978-1-5225-0256-2.ch015

Antonio, C. P., Simão, J. P., Alves, J. B., Bento da Silva, J., & Coelho dos Santos, A. (2017). Using 3D Virtual Worlds Integrated to Remote Experimentation in Sciences Teaching. In R. Alexandre Peixoto de Queirós & M. Pinto (Eds.), *Gamification-Based E-Learning Strategies for Computer Programming Education* (pp. 195–220). Hershey, PA: IGI Global. doi:10.4018/978-1-5225-1034-5.ch009

Asunka, S. (2016). Instructor Perceptions and Intentions to Use a Tablet PC for Mobile Learning in a Ghanaian University: An Exploratory Case Study. In L. Briz-Ponce, J. Juanes-Méndez, & F. García-Peñalvo (Eds.), *Handbook of Research on Mobile Devices and Applications in Higher Education Settings* (pp. 495–517). Hershey, PA: IGI Global. doi:10.4018/978-1-5225-0256-2.ch021

aus der Wieschen, M. V., Fischer, K., Kukliński, K., Jensen, L. C., & Savarimuthu, T. R. (2016). Multimodal Feedback in Human-Robot Interaction: An HCI-Informed Comparison of Feedback Modalities. In J. Rodrigues, P. Cardoso, J. Monteiro, & M. Figueiredo (Eds.), *Handbook of Research on Human-Computer Interfaces, Developments, and Applications* (pp. 135–161). Hershey, PA: IGI Global. doi:10.4018/978-1-5225-0435-1.ch006

Bachvarova, Y., & Bocconi, S. (2014). Games and Social Networks. In T. Connolly, T. Hainey, E. Boyle, G. Baxter, & P. Moreno-Ger (Eds.), *Psychology, Pedagogy, and Assessment in Serious Games* (pp. 204–219). Hershey, PA: IGI Global. doi:10.4018/978-1-4666-4773-2.ch010

Balog, A., & Pribeanu, C. (2016). An Extended Acceptance Model for Augmented Reality Educational Applications. In F. Neto, R. de Souza, & A. Gomes (Eds.), *Handbook of Research on 3-D Virtual Environments and Hypermedia for Ubiquitous Learning* (pp. 537–554). Hershey, PA: IGI Global. doi:10.4018/978-1-5225-0125-1.ch022

Bartoli, F. (2017). Coding CLIL Paths Inworld: Engaged Learning in Multilayered Environments. In G. Panconesi & M. Guida (Eds.), *Handbook of Research on Collaborative Teaching Practice in Virtual Learning Environments* (pp. 15–51). Hershey, PA: IGI Global. doi:10.4018/978-1-5225-2426-7.ch002

Ben Romdhane, D. B., Alkhatib, M. Y., & Legros, D. (2016). Memory, Cognition, and Multimedia: Role of the Importance of Organizing Knowledge in Memory with Hypermedia Help and Online Co-Understanding Systems. In F. Neto, R. de Souza, & A. Gomes (Eds.), *Handbook of Research on 3-D Virtual Environments and Hypermedia for Ubiquitous Learning* (pp. 318–337). Hershey, PA: IGI Global. doi:10.4018/978-1-5225-0125-1.ch013

Bermingham, N., & Prendergast, M. (2016). Bespoke Mobile Application Development: Facilitating Transition of Foundation Students to Higher Education. In L. Briz-Ponce, J. Juanes-Méndez, & F. García-Peñalvo (Eds.), *Handbook of Research on Mobile Devices and Applications in Higher Education Settings* (pp. 222–249). Hershey, PA: IGI Global. doi:10.4018/978-1-5225-0256-2.ch010

Blasko, D. G., Lum, H. C., White, M. M., & Drabik, H. B. (2014). Individual Differences in the Enjoyment and Effectiveness of Serious Games. In T. Connolly, T. Hainey, E. Boyle, G. Baxter, & P. Moreno-Ger (Eds.), *Psychology, Pedagogy, and Assessment in Serious Games* (pp. 153–174). Hershey, PA: IGI Global. doi:10.4018/978-1-4666-4773-2.ch008

Blummer, B., & Kenton, J. M. (2016). Academic Libraries' Mobile Initiatives and Research from 2010 to the Present: Identifying Themes in the Literature. In L. Briz-Ponce, J. Juanes-Méndez, & F. García-Peñalvo (Eds.), *Handbook of Research on Mobile Devices and Applications in Higher Education Settings* (pp. 118–139). Hershey, PA: IGI Global. doi:10.4018/978-1-5225-0256-2.ch006

Boniello, A., Paris, E., & Santoianni, F. (2017). Virtual Worlds in Geoscience Education: Learning Strategies and Learning 3D Environments. In G. Panconesi & M. Guida (Eds.), *Handbook of Research on Collaborative Teaching Practice in Virtual Learning Environments* (pp. 387–406). Hershey, PA: IGI Global. doi:10.4018/978-1-5225-2426-7.ch020

Borovikov, E., Zavorin, I., & Yershov, S. (2016). On Vision-Based Human-Centric Virtual Character Design: A Closer Look at the Real World from a Virtual One. In J. Turner, M. Nixon, U. Bernardet, & S. DiPaola (Eds.), *Integrating Cognitive Architectures into Virtual Character Design* (pp. 1–34). Hershey, PA: IGI Global. doi:10.4018/978-1-5225-0454-2.ch001

Borrelle, S. B., Frielick, S., Asshoff, R., & Leuzinger, S. (2016). The Global Change App: The Creative Transformation of Scientific Research. In L. Briz-Ponce, J. Juanes-Méndez, & F. García-Peñalvo (Eds.), *Handbook of Research on Mobile Devices and Applications in Higher Education Settings* (pp. 140–161). Hershey, PA: IGI Global. doi:10.4018/978-1-5225-0256-2.ch007

Borsani, M. (2017). Real Life and Second Life, Osmotic Membranes: Dynamics of Identity and Social Relationships in Second Life. In G. Panconesi & M. Guida (Eds.), *Handbook of Research on Collaborative Teaching Practice in Virtual Learning Environments* (pp. 258–274). Hershey, PA: IGI Global. doi:10.4018/978-1-5225-2426-7.ch014

Boyle, E. (2014). Psychological Aspects of Serious Games. In T. Connolly, T. Hainey, E. Boyle, G. Baxter, & P. Moreno-Ger (Eds.), *Psychology, Pedagogy, and Assessment in Serious Games* (pp. 1–18). Hershey, PA: IGI Global. doi:10.4018/978-1-4666-4773-2.ch001

Boyle, E., Terras, M. M., Ramsay, J., & Boyle, J. M. (2014). Executive Functions in Digital Games. In T. Connolly, T. Hainey, E. Boyle, G. Baxter, & P. Moreno-Ger (Eds.), *Psychology, Pedagogy, and Assessment in Serious Games* (pp. 19–46). Hershey, PA: IGI Global. doi:10.4018/978-1-4666-4773-2.ch002

Bringsjord, S., Licato, J., & Bringsjord, A. (2016). The Contemporary Craft of Creating Characters Meets Today's Cognitive Architectures: A Case Study in Expressivity. In J. Turner, M. Nixon, U. Bernardet, & S. DiPaola (Eds.), *Integrating Cognitive Architectures into Virtual Character Design* (pp. 151–180). Hershey, PA: IGI Global. doi:10.4018/978-1-5225-0454-2.ch006

Briz-Ponce, L., Juanes-Méndez, J. A., & García-Peñalvo, F. J. (2016). A Research Contribution to the Analysis of Mobile Devices in Higher Education from Medical Students' Point of View. In L. Briz-Ponce, J. Juanes-Méndez, & F. García-Peñalvo (Eds.), *Handbook of Research on Mobile Devices and Applications in Higher Education Settings* (pp. 196–221). Hershey, PA: IGI Global. doi:10.4018/978-1-5225-0256-2.ch009

Brkić, V. K., Putnik, G. D., Veljkovic, Z. A., & Shah, V. (2016). Interface for Distributed Remote User Controlled Manufacturing: Manufacturing and Education Sectors Led View. In J. Rodrigues, P. Cardoso, J. Monteiro, & M. Figueiredo (Eds.), *Handbook of Research on Human-Computer Interfaces, Developments, and Applications* (pp. 363–391). Hershey, PA: IGI Global. doi:10.4018/978-1-5225-0435-1.ch015

Busuttil, L., Gruppetta, C., & Camilleri, V. (2016). Minecraft Schooling: Digital Learning for Junior Years. In D. Russell & J. Laffey (Eds.), *Handbook of Research on Gaming Trends in P-12 Education* (pp. 481–501). Hershey, PA: IGI Global. doi:10.4018/978-1-4666-9629-7.ch023

Butcher, K. R., Runburg, M., & Altizer, R. (2017). Dino Lab: Designing and Developing an Educational Game for Critical Thinking. In R. Zheng & M. Gardner (Eds.), *Handbook of Research on Serious Games for Educational Applications* (pp. 115–148). Hershey, PA: IGI Global. doi:10.4018/978-1-5225-0513-6.ch006

Camilleri, V., de Freitas, S., Dunwell, I., & Montebello, M. (2017). Classroom Technology Acceptance for Teachers in 3D: A Case Study in PreVieW. In G. Panconesi & M. Guida (Eds.), *Handbook of Research on Collaborative Teaching Practice in Virtual Learning Environments* (pp. 221–240). Hershey, PA: IGI Global. doi:10.4018/978-1-5225-2426-7.ch012

Caponata, I., & Ferraro, A. P. (2017). Virtual Learning Environments for a New Teaching Methodology. In G. Panconesi & M. Guida (Eds.), *Handbook of Research on Collaborative Teaching Practice in Virtual Learning Environments* (pp. 76–91). Hershey, PA: IGI Global. doi:10.4018/978-1-5225-2426-7.ch004

Cardoso, P., Rodrigues, J., Monteiro, J., & Sousa, L. (2016). Using a Hands-Free System to Manage Common Devices in Constrained Conditions. In J. Rodrigues, P. Cardoso, J. Monteiro, & M. Figueiredo (Eds.), *Handbook of Research on Human-Computer Interfaces, Developments, and Applications* (pp. 73–98). Hershey, PA: IGI Global. doi:10.4018/978-1-5225-0435-1.ch004

Cardullo, V. M., & Clark, L. L. (2016). Universities' Point of View to Introduce Mobile Devices in their Classrooms: Redefining Education using a Common Mobile Platform – The Journey through Implementation. In L. Briz-Ponce, J. Juanes-Méndez, & F. García-Peñalvo (Eds.), *Handbook of Research on Mobile Devices and Applications in Higher Education Settings* (pp. 297–317). Hershey, PA: IGI Global. doi:10.4018/978-1-5225-0256-2.ch013

Choi, H., & Feng, J. (2017). Using Video Games to Improve Spatial Skills. In R. Zheng & M. Gardner (Eds.), *Handbook of Research on Serious Games for Educational Applications* (pp. 93–114). Hershey, PA: IGI Global. doi:10.4018/978-1-5225-0513-6.ch005

Cinto, T. (2016). Towards an Inclusive Walk-in Customer Service Facility. In J. Rodrigues, P. Cardoso, J. Monteiro, & M. Figueiredo (Eds.), *Handbook of Research on Human-Computer Interfaces, Developments, and Applications* (pp. 525–544). Hershey, PA: IGI Global. doi:10.4018/978-1-5225-0435-1.ch021

Cinto, T., Leite, H. M., Carvalho, S. N., Peixoto, C. S., & Arantes, D. S. (2016). 3D Virtual Learning Environments: An Avatar-Based Virtual Classes Platform. In F. Neto, R. de Souza, & A. Gomes (Eds.), *Handbook of Research on 3-D Virtual Environments and Hypermedia for Ubiquitous Learning* (pp. 54–86). Hershey, PA: IGI Global. doi:10.4018/978-1-5225-0125-1.ch003

Citarella, I. (2017). The Added Value of 3D World in Professional, Educational, and Individual Dynamics. In G. Panconesi & M. Guida (Eds.), *Handbook of Research on Collaborative Teaching Practice in Virtual Learning Environments* (pp. 275–297). Hershey, PA: IGI Global. doi:10.4018/978-1-5225-2426-7.ch015

Clark, D. B., Sengupta, P., & Virk, S. (2016). Disciplinarily-Integrated Games: Generalizing Across Domains and Model Types. In D. Russell & J. Laffey (Eds.), *Handbook of Research on Gaming Trends in P-12 Education* (pp. 178–194). Hershey, PA: IGI Global. doi:10.4018/978-1-4666-9629-7.ch009

Clinton, V., Cooper, J. L., Michaelis, J. E., Alibali, M. W., & Nathan, M. J. (2017). How Revisions to Mathematical Visuals Affect Cognition: Evidence from Eye Tracking. In C. Was, F. Sansosti, & B. Morris (Eds.), *Eye-Tracking Technology Applications in Educational Research* (pp. 195–218). Hershey, PA: IGI Global. doi:10.4018/978-1-5225-1005-5.ch010

Collins, P. K. (2017). Strategies to Support the Faculty Adoption of Technology for Student Success Initiatives. In T. Kidd & L. Morris Jr., (Eds.), *Handbook of Research on Instructional Systems and Educational Technology* (pp. 392–399). Hershey, PA: IGI Global. doi:10.4018/978-1-5225-2399-4.ch033

Connally, P. N., & Morris, L. R. (2017). The Impact of Emerging Technology on Leadership Development. In T. Kidd & L. Morris Jr., (Eds.), *Handbook of Research on Instructional Systems and Educational Technology* (pp. 400–408). Hershey, PA: IGI Global. doi:10.4018/978-1-5225-2399-4.ch034

Cook, A. E., & Wei, W. (2017). Using Eye Movements to Study Reading Processes: Methodological Considerations. In C. Was, F. Sansosti, & B. Morris (Eds.), *Eye-Tracking Technology Applications in Educational Research* (pp. 27–47). Hershey, PA: IGI Global. doi:10.4018/978-1-5225-1005-5.ch002

Corradini, A., & Mehta, M. (2016). A Graphical Tool for the Creation of Behaviors in Virtual Worlds. In J. Turner, M. Nixon, U. Bernardet, & S. DiPaola (Eds.), *Integrating Cognitive Architectures into Virtual Character Design* (pp. 65–93). Hershey, PA: IGI Global. doi:10.4018/978-1-5225-0454-2.ch003

Cram, A., Lowe, R., & Lumkin, K. (2014). Assessing Spatial Design in Virtual Environments. In S. Kennedy-Clark, K. Everett, & P. Wheeler (Eds.), *Cases on the Assessment of Scenario and Game-Based Virtual Worlds in Higher Education* (pp. 74–123). Hershey, PA: IGI Global. doi:10.4018/978-1-4666-4470-0.ch003

DaCosta, B., & Seok, S. (2017). Factors That Explain Adolescent and Young Adult Mobile Game Play, Part 1: A Quantitative Examination of the Characteristics Describing the Casual Player. In R. Zheng & M. Gardner (Eds.), *Handbook of Research on Serious Games for Educational Applications* (pp. 320–339). Hershey, PA: IGI Global. doi:10.4018/978-1-5225-0513-6.ch015

DaCosta, B., & Seok, S. (2017). Factors That Explain Adolescent and Young Adult Mobile Game Play, Part 2: A Quantitative Examination of the Casual Player in the Context of Age and Gender. In R. Zheng & M. Gardner (Eds.), *Handbook of Research on Serious Games for Educational Applications* (pp. 340–365). Hershey, PA: IGI Global. doi:10.4018/978-1-5225-0513-6.ch016

Dass, S., & Dabbagh, N. (2016). Faculty Adoption of 3D Avatar-Based Virtual World Learning Environments: An Exploratory Case Study. In F. Neto, R. de Souza, & A. Gomes (Eds.), *Handbook of Research on 3-D Virtual Environments and Hypermedia for Ubiquitous Learning* (pp. 262–296). Hershey, PA: IGI Global. doi:10.4018/978-1-5225-0125-1.ch011

Davis, D., & Stewart, S. (2014). Factors Shaping Assessment Design in the Virtual Environment: A Case Study of Midwifery. In S. Kennedy-Clark, K. Everett, & P. Wheeler (Eds.), *Cases on the Assessment of Scenario and Game-Based Virtual Worlds in Higher Education* (pp. 1–35). Hershey, PA: IGI Global. doi:10.4018/978-1-4666-4470-0.ch001

Davis, L., Dawson, M., & Omar, M. (2016). Systems Engineering Concepts with Aid of Virtual Worlds and Open Source Software: Using Technology to Develop Learning Objects and Simulation Environments. In F. Neto, R. de Souza, & A. Gomes (Eds.), *Handbook of Research on 3-D Virtual Environments and Hypermedia for Ubiquitous Learning* (pp. 483–509). Hershey, PA: IGI Global. doi:10.4018/978-1-5225-0125-1.ch020

De Pasquale, D., Wood, E., Gottardo, A., Jones, J. A., Kaplan, R., & DeMarco, A. (2017). Tracking Children's Interactions with Traditional Text and Computer-Based Early Literacy Media. In C. Was, F. Sansosti, & B. Morris (Eds.), *Eye-Tracking Technology Applications in Educational Research* (pp. 107–121). Hershey, PA: IGI Global. doi:10.4018/978-1-5225-1005-5.ch006

Demir, K. (2017). Mobile Learning. In T. Kidd & L. Morris Jr., (Eds.), *Handbook of Research on Instructional Systems and Educational Technology* (pp. 291–298). Hershey, PA: IGI Global. doi:10.4018/978-1-5225-2399-4.ch025

Demirbilek, M., & Siauciune, V. K. (2017). Foreign Language Teaching Practices in 3D Multi-User Virtual Immersive Learning Environments. In G. Panconesi & M. Guida (Eds.), *Handbook of Research on Collaborative Teaching Practice in Virtual Learning Environments* (pp. 129–145). Hershey, PA: IGI Global. doi:10.4018/978-1-5225-2426-7.ch007

Demiroz, E. (2016). Kinesthetic Gaming, Cognition, and Learning: Implications for P-12 Education. In D. Russell & J. Laffey (Eds.), *Handbook of Research on Gaming Trends in P-12 Education* (pp. 124–150). Hershey, PA: IGI Global. doi:10.4018/978-1-4666-9629-7.ch006

Desjarlais, M. (2017). The Use of Eye-gaze to Understand Multimedia Learning. In C. Was, F. Sansosti, & B. Morris (Eds.), *Eye-Tracking Technology Applications in Educational Research* (pp. 122–142). Hershey, PA: IGI Global. doi:10.4018/978-1-5225-1005-5.ch007

Dreon, O., & Szczyrbak, G. (2017). Level Up: Multiple Player Professional Development. In R. Zheng & M. Gardner (Eds.), *Handbook of Research on Serious Games for Educational Applications* (pp. 366–379). Hershey, PA: IGI Global. doi:10.4018/978-1-5225-0513-6.ch017

Dubbels, B. R. (2017). Gamification Transformed: Gamification Should Deliver the Best Parts of Game Experiences, Not Just Experiences of Game Parts. In B. Dubbels (Ed.), *Transforming Gaming and Computer Simulation Technologies across Industries* (pp. 17–47). Hershey, PA: IGI Global. doi:10.4018/978-1-5225-1817-4.ch002

DuQuette, J. L. (2017). Lessons from Cypris Chat: Revisiting Virtual Communities as Communities. In G. Panconesi & M. Guida (Eds.), *Handbook of Research on Collaborative Teaching Practice in Virtual Learning Environments* (pp. 299–316). Hershey, PA: IGI Global. doi:10.4018/978-1-5225-2426-7.ch016

Ebner, M., & Spot, N. (2016). Game-Based Learning with the Leap Motion Controller. In D. Russell & J. Laffey (Eds.), *Handbook of Research on Gaming Trends in P-12 Education* (pp. 555–565). Hershey, PA: IGI Global. doi:10.4018/978-1-4666-9629-7.ch026

Fabbri, F. L. (2017). Popularization of Science in 3D Virtual Environments With “Scienza on the Road”. In G. Panconesi & M. Guida (Eds.), *Handbook of Research on Collaborative Teaching Practice in Virtual Learning Environments* (pp. 517–548). Hershey, PA: IGI Global. doi:10.4018/978-1-5225-2426-7.ch026

Farmeschi, N. (2017). Dante Project: Experience the Immersive Virtual Worlds. In G. Panconesi & M. Guida (Eds.), *Handbook of Research on Collaborative Teaching Practice in Virtual Learning Environments* (pp. 493–516). Hershey, PA: IGI Global. doi:10.4018/978-1-5225-2426-7.ch025

Ficagna, T. V., & Raabe, A. (2016). Impact Assessment of Affective Virtual Characters in a Virtual Environment. In F. Neto, R. de Souza, & A. Gomes (Eds.), *Handbook of Research on 3-D Virtual Environments and Hypermedia for Ubiquitous Learning* (pp. 220–242). Hershey, PA: IGI Global. doi:10.4018/978-1-5225-0125-1.ch009

Finco, G. (2017). Energy in 3D: Designing the City of the Future. In G. Panconesi & M. Guida (Eds.), *Handbook of Research on Collaborative Teaching Practice in Virtual Learning Environments* (pp. 444–463). Hershey, PA: IGI Global. doi:10.4018/978-1-5225-2426-7.ch023

Fisher, K. J., Nichols, T., Isbister, K., & Fuller, T. (2017). Quantifying "Magic": Creating Good Player Experiences on Xbox Kinect. In B. Dubbels (Ed.), *Transforming Gaming and Computer Simulation Technologies across Industries* (pp. 1–16). Hershey, PA: IGI Global. doi:10.4018/978-1-5225-1817-4.ch001

Floyd, K. K., & Shambaugh, N. (2017). Instructional Design for Simulations in Special Education Virtual Learning Spaces. In T. Kidd & L. Morris Jr., (Eds.), *Handbook of Research on Instructional Systems and Educational Technology* (pp. 202–215). Hershey, PA: IGI Global. doi:10.4018/978-1-5225-2399-4.ch018

Folk, J. R., & Eskenazi, M. A. (2017). Eye Movement Behavior and Individual Differences in Word Identification During Reading. In C. Was, F. Sansosti, & B. Morris (Eds.), *Eye-Tracking Technology Applications in Educational Research* (pp. 66–87). Hershey, PA: IGI Global. doi:10.4018/978-1-5225-1005-5.ch004

Forgette, J., & Katchabaw, M. (2016). Learned Behavior: Enabling Believable Virtual Characters through Reinforcement. In J. Turner, M. Nixon, U. Bernardet, & S. DiPaola (Eds.), *Integrating Cognitive Architectures into Virtual Character Design* (pp. 94–123). Hershey, PA: IGI Global. doi:10.4018/978-1-5225-0454-2.ch004

Frade, R. V., Neto, F. M., & Castro de Souza, R. (2016). A Three-Dimensional Environment of Personalized Recommendation of Learning Objects to Support Ubiquitous Learning. In F. Neto, R. de Souza, & A. Gomes (Eds.), *Handbook of Research on 3-D Virtual Environments and Hypermedia for Ubiquitous Learning* (pp. 555–580). Hershey, PA: IGI Global. doi:10.4018/978-1-5225-0125-1.ch023

Fresno, J., Ortega-Arranz, H., Ortega-Arranz, A., Gonzalez-Escribano, A., & Llanos, D. R. (2017). Applying Gamification in a Parallel Programming Course. In R. Alexandre Peixoto de Queirós & M. Pinto (Eds.), *Gamification-Based E-Learning Strategies for Computer Programming Education* (pp. 106–130). Hershey, PA: IGI Global. doi:10.4018/978-1-5225-1034-5.ch006

Gallegos, B., Kepple, M. T., & Bukaty, C. A. (2016). Using Video Gameplay to Measure Achievement for Students with Disabilities: A New Perspective to Grading and Achievement Reporting. In D. Russell & J. Laffey (Eds.), *Handbook of Research on Gaming Trends in P-12 Education* (pp. 326–352). Hershey, PA: IGI Global. doi:10.4018/978-1-4666-9629-7.ch016

Gardner, M. K., & Strayer, D. L. (2017). What Cognitive Psychology Can Tell Us About Educational Computer Games. In R. Zheng & M. Gardner (Eds.), *Handbook of Research on Serious Games for Educational Applications* (pp. 1–18). Hershey, PA: IGI Global. doi:10.4018/978-1-5225-0513-6.ch001

Ge, X., & Ifenthaler, D. (2017). Designing Engaging Educational Games and Assessing Engagement in Game-Based Learning. In R. Zheng & M. Gardner (Eds.), *Handbook of Research on Serious Games for Educational Applications* (pp. 253–270). Hershey, PA: IGI Global. doi:10.4018/978-1-5225-0513-6.ch012

Giannetti, L. (2017). Geometry and Pastry Making/Tesselations and Decorations: Teaching Experience Between Real and Virtual. In G. Panconesi & M. Guida (Eds.), *Handbook of Research on Collaborative Teaching Practice in Virtual Learning Environments* (pp. 464–492). Hershey, PA: IGI Global. doi:10.4018/978-1-5225-2426-7.ch024

Gil, M., & Otero, B. (2017). Learning Engineering Skills through Creativity and Collaboration: A Game-Based Proposal. In R. Alexandre Peixoto de Queirós & M. Pinto (Eds.), *Gamification-Based E-Learning Strategies for Computer Programming Education* (pp. 14–29). Hershey, PA: IGI Global. doi:10.4018/978-1-5225-1034-5.ch002

Gomes, A. D., Santos, Á. N., Páris, C. P., & Martins, N. C. (2017). Playing with Programming: A Serious Game to Start Programming. In R. Alexandre Peixoto de Queirós & M. Pinto (Eds.), *Gamification-Based E-Learning Strategies for Computer Programming Education* (pp. 261–277). Hershey, PA: IGI Global. doi:10.4018/978-1-5225-1034-5.ch012

Gromala, D., Tong, X., Shaw, C., & Jin, W. (2016). Immersive Virtual Reality as a Non-Pharmacological Analgesic for Pain Management: Pain Distraction and Pain Self-modulation. In J. Rodrigues, P. Cardoso, J. Monteiro, & M. Figueiredo (Eds.), *Handbook of Research on Human-Computer Interfaces, Developments, and Applications* (pp. 478–500). Hershey, PA: IGI Global. doi:10.4018/978-1-5225-0435-1.ch019

Gros, B. (2017). Game Dimensions and Pedagogical Dimension in Serious Games. In R. Zheng & M. Gardner (Eds.), *Handbook of Research on Serious Games for Educational Applications* (pp. 402–417). Hershey, PA: IGI Global. doi:10.4018/978-1-5225-0513-6.ch019

Gül, L. F., Gu, N., Kim, M. J., & Wang, X. (2014). Methods for Assessing 3D Virtual Worlds in Design Education. In S. Kennedy-Clark, K. Everett, & P. Wheeler (Eds.), *Cases on the Assessment of Scenario and Game-Based Virtual Worlds in Higher Education* (pp. 152–174). Hershey, PA: IGI Global. doi:10.4018/978-1-4666-4470-0.ch005

Güngör, A. S., Çadırcı, T. O., & Köse, Ş. G. (2016). Advergaming – How Does Cognitive Overload Effect Brand Recall?: Differences between In-Game Advertising (IGA) and Advergames. In J. Rodrigues, P. Cardoso, J. Monteiro, & M. Figueiredo (Eds.), *Handbook of Research on Human-Computer Interfaces, Developments, and Applications* (pp. 501–524). Hershey, PA: IGI Global. doi:10.4018/978-1-5225-0435-1.ch020

Güzeldereli, E. A., Kose, U., & Tüfekci, A. (2017). Human Performance Technology and the Effects on Web-Based Instruction Performance Efficiency. In T. Kidd & L. Morris Jr., (Eds.), *Handbook of Research on Instructional Systems and Educational Technology* (pp. 89–103). Hershey, PA: IGI Global. doi:10.4018/978-1-5225-2399-4.ch009

Hacker, D. J. (2017). The Role of Metacognition in Learning via Serious Games. In R. Zheng & M. Gardner (Eds.), *Handbook of Research on Serious Games for Educational Applications* (pp. 19–40). Hershey, PA: IGI Global. doi:10.4018/978-1-5225-0513-6.ch002

Hadjistassou, S. K., & Molka-Danielsen, J. (2016). Designing Alien Mysteries in Chatterdale: An Instructor's Perspective. In D. Russell & J. Laffey (Eds.), *Handbook of Research on Gaming Trends in P-12 Education* (pp. 222–236). Hershey, PA: IGI Global. doi:10.4018/978-1-4666-9629-7.ch011

Hainey, T., Connolly, T. M., Chaudy, Y., Boyle, E., Beeby, R., & Soflano, M. (2014). Assessment Integration in Serious Games. In T. Connolly, T. Hainey, E. Boyle, G. Baxter, & P. Moreno-Ger (Eds.), *Psychology, Pedagogy, and Assessment in Serious Games* (pp. 317–341). Hershey, PA: IGI Global. doi:10.4018/978-1-4666-4773-2.ch015

Hainey, T., Soflano, M., & Connolly, T. M. (2014). A Randomised Controlled Trial to Evaluate Learning Effectiveness Using an Adaptive Serious Game to Teach SQL at Higher Education Level. In T. Connolly, T. Hainey, E. Boyle, G. Baxter, & P. Moreno-Ger (Eds.), *Psychology, Pedagogy, and Assessment in Serious Games* (pp. 270–291). Hershey, PA: IGI Global. doi:10.4018/978-1-4666-4773-2.ch013

Hassani, K., & Lee, W. (2016). A Universal Architecture for Migrating Cognitive Agents: A Case Study on Automatic Animation Generation. In J. Turner, M. Nixon, U. Bernardet, & S. DiPaola (Eds.), *Integrating Cognitive Architectures into Virtual Character Design* (pp. 238–265). Hershey, PA: IGI Global. doi:10.4018/978-1-5225-0454-2.ch009

Hauge, J. B., Boyle, E., Mayer, I., Nadolski, R., Riedel, J. C., Moreno-Ger, P., & Ritchie, J. et al. (2014). Study Design and Data Gathering Guide for Serious Games' Evaluation. In T. Connolly, T. Hainey, E. Boyle, G. Baxter, & P. Moreno-Ger (Eds.), *Psychology, Pedagogy, and Assessment in Serious Games* (pp. 394–419). Hershey, PA: IGI Global. doi:10.4018/978-1-4666-4773-2.ch018

Havard, B., & Podsiad, M. (2017). Wearable Computers. In T. Kidd & L. Morris Jr., (Eds.), *Handbook of Research on Instructional Systems and Educational Technology* (pp. 356–365). Hershey, PA: IGI Global. doi:10.4018/978-1-5225-2399-4.ch030

Herpich, F., Nunes, F. B., Voss, G. B., & Medina, R. D. (2016). Three-Dimensional Virtual Environment and NPC: A Perspective about Intelligent Agents Ubiquitous. In F. Neto, R. de Souza, & A. Gomes (Eds.), *Handbook of Research on 3-D Virtual Environments and Hypermedia for Ubiquitous Learning* (pp. 510–536). Hershey, PA: IGI Global. doi:10.4018/978-1-5225-0125-1.ch021

Herrera-Bernal, J., Ramírez-Hernández, D. D., & Ramírez-Montoya, M. (2016). Applied Competences for Students by Using M-Learning Devices in Higher Education: Knowledge, Skills, and Attitudes. In L. Briz-Ponce, J. Juanes-Méndez, & F. García-Peñalvo (Eds.), *Handbook of Research on Mobile Devices and Applications in Higher Education Settings* (pp. 453–476). Hershey, PA: IGI Global. doi:10.4018/978-1-5225-0256-2.ch019

Hill, V. J. (2016). Information Literacy in Virtual Environments: Changing Needs of P-12 Learners. In D. Russell & J. Laffey (Eds.), *Handbook of Research on Gaming Trends in P-12 Education* (pp. 165–177). Hershey, PA: IGI Global. doi:10.4018/978-1-4666-9629-7.ch008

Holloway, A. (2017). Teaching Childbirth Support Techniques Using the Prepared Partner and Digital Birth: The Design and Development of Games for Dads-To-Be. In B. Dubbels (Ed.), *Transforming Gaming and Computer Simulation Technologies across Industries* (pp. 154–182). Hershey, PA: IGI Global. doi:10.4018/978-1-5225-1817-4.ch009

Hricko, M. (2017). Personal Learning Environments. In T. Kidd & L. Morris Jr., (Eds.), *Handbook of Research on Instructional Systems and Educational Technology* (pp. 236–248). Hershey, PA: IGI Global. doi:10.4018/978-1-5225-2399-4.ch021

Huber, E., & Blount, Y. (2014). Second Life for Assessing Postgraduate Learning: Student Perspectives. In S. Kennedy-Clark, K. Everett, & P. Wheeler (Eds.), *Cases on the Assessment of Scenario and Game-Based Virtual Worlds in Higher Education* (pp. 36–73). Hershey, PA: IGI Global. doi:10.4018/978-1-4666-4470-0.ch002

Hung, Y., Chang, R. I., & Lin, C. F. (2016). Survey of Students' Perception and the Influence of Learning Style Preferences on Learner's Intention to Use 3D Visualization Instruction: Learning Style and 3D Visualization Instruction. In F. Neto, R. de Souza, & A. Gomes (Eds.), *Handbook of Research on 3-D Virtual Environments and Hypermedia for Ubiquitous Learning* (pp. 297–316). Hershey, PA: IGI Global. doi:10.4018/978-1-5225-0125-1.ch012

Ibañez, R., Soria, A., Teyseyre, A. R., Berdun, L., & Campo, M. R. (2016). A Comparative Study of Machine Learning Techniques for Gesture Recognition Using Kinect. In J. Rodrigues, P. Cardoso, J. Monteiro, & M. Figueiredo (Eds.), *Handbook of Research on Human-Computer Interfaces, Developments, and Applications* (pp. 1–22). Hershey, PA: IGI Global. doi:10.4018/978-1-5225-0435-1.ch001

Islam, M. N., & Tétard, F. (2016). Using a Design Science Research Approach to Develop a HCI Artifact for Designing Web Interfaces: A Case Study. In J. Rodrigues, P. Cardoso, J. Monteiro, & M. Figueiredo (Eds.), *Handbook of Research on Human-Computer Interfaces, Developments, and Applications* (pp. 304–320). Hershey, PA: IGI Global. doi:10.4018/978-1-5225-0435-1.ch012

Jaklin, N., & Geraerts, R. (2016). Navigating Through Virtual Worlds: From Single Characters to Large Crowds. In D. Russell & J. Laffey (Eds.), *Handbook of Research on Gaming Trends in P-12 Education* (pp. 527–554). Hershey, PA: IGI Global. doi:10.4018/978-1-4666-9629-7.ch025

Jantjies, M., & Joy, M. (2017). Teaching Through Mobile Technology: A Reflection From High School Studies in South Africa. In T. Kidd & L. Morris Jr., (Eds.), *Handbook of Research on Instructional Systems and Educational Technology* (pp. 299–312). Hershey, PA: IGI Global. doi:10.4018/978-1-5225-2399-4.ch026

Jones, G., & Alba, A. D. (2016). Reviewing the Effectiveness and Learning Outcomes of a 3D Virtual Museum: A Pilot Study. In F. Neto, R. de Souza, & A. Gomes (Eds.), *Handbook of Research on 3-D Virtual Environments and Hypermedia for Ubiquitous Learning* (pp. 168–191). Hershey, PA: IGI Global. doi:10.4018/978-1-5225-0125-1.ch007

Kalaian, S. A. (2017). Distance and Online Learning. In T. Kidd & L. Morris Jr., (Eds.), *Handbook of Research on Instructional Systems and Educational Technology* (pp. 23–36). Hershey, PA: IGI Global. doi:10.4018/978-1-5225-2399-4.ch003

Kaplan, E., Levari, T., & Snedeker, J. (2017). Eye Tracking and Spoken Language Comprehension. In C. Was, F. Sansosti, & B. Morris (Eds.), *Eye-Tracking Technology Applications in Educational Research* (pp. 88–105). Hershey, PA: IGI Global. doi:10.4018/978-1-5225-1005-5.ch005

Kari, T. (2017). Promoting Physical Activity and Fitness with Exergames: Updated Systematic Review of Systematic Reviews. In B. Dubbels (Ed.), *Transforming Gaming and Computer Simulation Technologies across Industries* (pp. 225–245). Hershey, PA: IGI Global. doi:10.4018/978-1-5225-1817-4.ch013

Kasemsap, K. (2017). Mastering Distance Education in the Digital Age. In T. Kidd & L. Morris Jr., (Eds.), *Handbook of Research on Instructional Systems and Educational Technology* (pp. 11–22). Hershey, PA: IGI Global. doi:10.4018/978-1-5225-2399-4.ch002

Kasemsap, K. (2017). Mastering Educational Computer Games, Educational Video Games, and Serious Games in the Digital Age. In R. Alexandre Peixoto de Queirós & M. Pinto (Eds.), *Gamification-Based E-Learning Strategies for Computer Programming Education* (pp. 30–52). Hershey, PA: IGI Global. doi:10.4018/978-1-5225-1034-5.ch003

Kasemsap, K. (2017). Trends and Issues With Massive Open Online Courses. In T. Kidd & L. Morris Jr., (Eds.), *Handbook of Research on Instructional Systems and Educational Technology* (pp. 116–127). Hershey, PA: IGI Global. doi:10.4018/978-1-5225-2399-4.ch011

Katz, D., Goldberg, A., Khanal, P., Kahol, K., & DeMaria, S. (2017). Using Serious Gaming to Improve the Safety of Central Venous Catheter Placement: A Post-Mortem Analysis. In B. Dubbels (Ed.), *Transforming Gaming and Computer Simulation Technologies across Industries* (pp. 120–129). Hershey, PA: IGI Global. doi:10.4018/978-1-5225-1817-4.ch007

Kaufman, G., & Flanagan, M. (2017). Digital Divide: Comparing the Impact of Digital and Non-Digital Platforms on Player Behaviors and Game Impact. In B. Dubbels (Ed.), *Transforming Gaming and Computer Simulation Technologies across Industries* (pp. 94–101). Hershey, PA: IGI Global. doi:10.4018/978-1-5225-1817-4.ch005

Ke, F. (2017). Designing Intrinsic Integration of Learning and Gaming Actions in a 3D Architecture Game. In R. Zheng & M. Gardner (Eds.), *Handbook of Research on Serious Games for Educational Applications* (pp. 234–252). Hershey, PA: IGI Global. doi:10.4018/978-1-5225-0513-6.ch011

Kennedy-Clark, S., & Wheeler, P. (2014). Using Discourse Analysis to Assess Student Problem-Solving in a Virtual World. In S. Kennedy-Clark, K. Everett, & P. Wheeler (Eds.), *Cases on the Assessment of Scenario and Game-Based Virtual Worlds in Higher Education* (pp. 211–253). Hershey, PA: IGI Global. doi:10.4018/978-1-4666-4470-0.ch007

Kenttälä, V., Rousi, R., & Kankaanranta, M. (2017). Towards the Learning Experience Technology Usability Framework. In T. Kidd & L. Morris Jr., (Eds.), *Handbook of Research on Instructional Systems and Educational Technology* (pp. 128–140). Hershey, PA: IGI Global. doi:10.4018/978-1-5225-2399-4.ch012

Kert, S. B., & Erkoç, M. F. (2017). Game-Based Approaches, Gamification, and Programming Language Training. In R. Alexandre Peixoto de Queirós & M. Pinto (Eds.), *Gamification-Based E-Learning Strategies for Computer Programming Education* (pp. 131–151). Hershey, PA: IGI Global. doi:10.4018/978-1-5225-1034-5.ch007

Kidd, T. T. (2017). The Rhetoric of Fear: Voices and Stories Told of Faculty Who Engage in Online Teaching. In T. Kidd & L. Morris Jr., (Eds.), *Handbook of Research on Instructional Systems and Educational Technology* (pp. 69–77). Hershey, PA: IGI Global. doi:10.4018/978-1-5225-2399-4

Kidd, T. T., & Hodges, J. R. (2017). Social Media, Mobile Technology, and New Learning Opportunities: Implications for Social Justice and Educational Spaces in Schools. In T. Kidd & L. Morris Jr., (Eds.), *Handbook of Research on Instructional Systems and Educational Technology* (pp. 274–290). Hershey, PA: IGI Global. doi:10.4018/978-1-5225-2399-4

Kirginas, S., & Gouscos, D. (2016). Exploring the Impact of Free-Form and Structured Digital Games on the Player Experience of Kindergarten and Primary School Students. In D. Russell & J. Laffey (Eds.), *Handbook of Research on Gaming Trends in P-12 Education* (pp. 394–420). Hershey, PA: IGI Global. doi:10.4018/978-1-4666-9629-7.ch019

Kirschner, D. (2017). The Development of a Gamified System for Health Activism as a Graduate Student Project. In R. Zheng & M. Gardner (Eds.), *Handbook of Research on Serious Games for Educational Applications* (pp. 380–401). Hershey, PA: IGI Global. doi:10.4018/978-1-5225-0513-6.ch018

Koc, S., & Boboc, M. (2017). Structuring Online Instruction by Dynamic Design, Delivery, and Assessment. In T. Kidd & L. Morris Jr., (Eds.), *Handbook of Research on Instructional Systems and Educational Technology* (pp. 78–88). Hershey, PA: IGI Global. doi:10.4018/978-1-5225-2399-4.ch008

Konstantinidis, A. (2017). Using SmallWorlds to Enhance Social Presence, Group Cohesion, and Group Identity in an Online Postgraduate Course. In G. Panconesi & M. Guida (Eds.), *Handbook of Research on Collaborative Teaching Practice in Virtual Learning Environments* (pp. 92–110). Hershey, PA: IGI Global. doi:10.4018/978-1-5225-2426-7.ch005

Laffey, J. M., Sadler, T. D., Goggins, S. P., Griffin, J., & Babiuch, R. N. (2016). Mission HydroSci: Distance Learning through Game-Based 3D Virtual Learning Environments. In D. Russell & J. Laffey (Eds.), *Handbook of Research on Gaming Trends in P-12 Education* (pp. 421–441). Hershey, PA: IGI Global. doi:10.4018/978-1-4666-9629-7.ch020

Lai, Y., Leung, P. P., Wong, T., & Yuen, T. R. (2016). Exploring the Use of Mobile Devices and Applications to Support Field Experience in Teacher Education Courses. In L. Briz-Ponce, J. Juanes-Méndez, & F. García-Peñalvo (Eds.), *Handbook of Research on Mobile Devices and Applications in Higher Education Settings* (pp. 477–494). Hershey, PA: IGI Global. doi:10.4018/978-1-5225-0256-2.ch020

Lamonaca, S. (2017). Art and Didactics in Virtual Worlds: How Technology Can Activate Social Skills. In G. Panconesi & M. Guida (Eds.), *Handbook of Research on Collaborative Teaching Practice in Virtual Learning Environments* (pp. 427–443). Hershey, PA: IGI Global. doi:10.4018/978-1-5225-2426-7.ch022

Lamonaca, S. (2017). Communication 2.0 at School: A Way to Connect Teachers and Students. In G. Panconesi & M. Guida (Eds.), *Handbook of Research on Collaborative Teaching Practice in Virtual Learning Environments* (pp. 374–385). Hershey, PA: IGI Global. doi:10.4018/978-1-5225-2426-7.ch019

Landers, R. N., & Callan, R. C. (2017). An Experiment on Anonymity and Multi-User Virtual Environments: Manipulating Identity to Increase Learning. In B. Dubbels (Ed.), *Transforming Gaming and Computer Simulation Technologies across Industries* (pp. 80–93). Hershey, PA: IGI Global. doi:10.4018/978-1-5225-1817-4.ch004

Lee, J., & Liu, M. (2017). Design of Fantasy and Their Effect on Learning and Engagement in a Serious Game. In R. Zheng & M. Gardner (Eds.), *Handbook of Research on Serious Games for Educational Applications* (pp. 197–216). Hershey, PA: IGI Global. doi:10.4018/978-1-5225-0513-6.ch009

Lemoine, P. A., Hackett, P. T., & Richardson, M. D. (2016). The Impact of Social Media on Instruction in Higher Education. In L. Briz-Ponce, J. Juanes-Méndez, & F. García-Peñalvo (Eds.), *Handbook of Research on Mobile Devices and Applications in Higher Education Settings* (pp. 373–401). Hershey, PA: IGI Global. doi:10.4018/978-1-5225-0256-2.ch016

Lima da Costa, A. A., Neto, F. M., Sombra, E. L., Moreira, J. D., Castro de Souza, R., & Dutra, J. G. (2016). Content Personalized Recommendation Engine to Support an Informal Learning Environment in the Health Context. In F. Neto, R. de Souza, & A. Gomes (Eds.), *Handbook of Research on 3-D Virtual Environments and Hypermedia for Ubiquitous Learning* (pp. 451–482). Hershey, PA: IGI Global. doi:10.4018/978-1-5225-0125-1.ch019

Lindsay, W. K. (2017). A Summary of Four Key Issues Affecting Distance Education. In T. Kidd & L. Morris Jr., (Eds.), *Handbook of Research on Instructional Systems and Educational Technology* (pp. 37–44). Hershey, PA: IGI Global. doi:10.4018/978-1-5225-2399-4.ch004

Litinski, V. (2017). Is Artificial Intelligence (AI) Friend or Foe to Patients in Healthcare?: On Virtues of Dynamic Consent – How to Build a Business Case for Digital Health Applications. In B. Dubbels (Ed.), *Transforming Gaming and Computer Simulation Technologies across Industries* (pp. 246–257). Hershey, PA: IGI Global. doi:10.4018/978-1-5225-1817-4.ch014

Liu, M., Horton, L., Kang, J., Kimmons, R. M., & Lee, J. (2017). Making Learning Fun: An Investigation of Using a Ludic Simulation for Middle School Space Science. In B. Dubbels (Ed.), *Transforming Gaming and Computer Simulation Technologies across Industries* (pp. 130–152). Hershey, PA: IGI Global. doi:10.4018/978-1-5225-1817-4.ch008

Liu, M., Su, S., Liu, S., Harron, J., Fickert, C., & Sherman, B. (2016). Exploring 3D Immersive and Interactive Technology for Designing Educational Learning Experiences. In F. Neto, R. de Souza, & A. Gomes (Eds.), *Handbook of Research on 3-D Virtual Environments and Hypermedia for Ubiquitous Learning* (pp. 243–261). Hershey, PA: IGI Global. doi:10.4018/978-1-5225-0125-1.ch010

Liu, Y. (2017). Gender Difference in Perception and Use of Social Media Tools. In T. Kidd & L. Morris Jr., (Eds.), *Handbook of Research on Instructional Systems and Educational Technology* (pp. 249–262). Hershey, PA: IGI Global. doi:10.4018/978-1-5225-2399-4.ch022

Luo, L., Kiewra, K. A., Peteranetz, M. S., & Flanigan, A. E. (2017). Using Eye-Tracking Technology to Understand How Graphic Organizers Aid Student Learning. In C. Was, F. Sansosti, & B. Morris (Eds.), *Eye-Tracking Technology Applications in Educational Research* (pp. 220–238). Hershey, PA: IGI Global. doi:10.4018/978-1-5225-1005-5.ch011

MacCallum, K., Day, S., Skelton, D., & Verhaart, M. (2017). Mobile Affordances and Learning Theories in Supporting and Enhancing Learning. *International Journal of Mobile and Blended Learning*, *9*(2), 61–73. doi:10.4018/IJMBL.2017040104

Mahmud, U. (2016). Organizing Contextual Data in Context Aware Systems: A Review. In J. Rodrigues, P. Cardoso, J. Monteiro, & M. Figueiredo (Eds.), *Handbook of Research on Human-Computer Interfaces, Developments, and Applications* (pp. 273–303). Hershey, PA: IGI Global. doi:10.4018/978-1-5225-0435-1.ch011

Manzoor, A. (2017). Blended Learning and Distance Education: Enablers and Barriers. In T. Kidd & L. Morris Jr., (Eds.), *Handbook of Research on Instructional Systems and Educational Technology* (pp. 45–57). Hershey, PA: IGI Global. doi:10.4018/978-1-5225-2399-4.ch005

Manzoor, A. (2017). Blended Learning in Higher Education: A Developing Country Perspective. In T. Kidd & L. Morris Jr., (Eds.), *Handbook of Research on Instructional Systems and Educational Technology* (pp. 58–68). Hershey, PA: IGI Global. doi:10.4018/978-1-5225-2399-4.ch006

Manzoor, A. (2017). Pedagogical Potential of Virtual Worlds: Challenges and Opportunities. In G. Panconesi & M. Guida (Eds.), *Handbook of Research on Collaborative Teaching Practice in Virtual Learning Environments* (pp. 146–161). Hershey, PA: IGI Global. doi:10.4018/978-1-5225-2426-7.ch008

Martin, N. L., & Soares, A. (2017). The Introductory Programming Course: A Game Design Approach for the E-Learning Environment. In R. Alexandre Peixoto de Queirós & M. Pinto (Eds.), *Gamification-Based E-Learning Strategies for Computer Programming Education* (pp. 79–104). Hershey, PA: IGI Global. doi:10.4018/978-1-5225-1034-5.ch005

Martinez-Garza, M. M., Clark, D. B., Killingsworth, S. S., & Adams, D. M. (2016). Beyond Fun: Pintrich, Motivation to Learn, and Games for Learning. In D. Russell & J. Laffey (Eds.), *Handbook of Research on Gaming Trends in P-12 Education* (pp. 1–32). Hershey, PA: IGI Global. doi:10.4018/978-1-4666-9629-7.ch001

Mayer, I., Bekebrede, G., Warmelink, H., & Zhou, Q. (2014). A Brief Methodology for Researching and Evaluating Serious Games and Game-Based Learning. In T. Connolly, T. Hainey, E. Boyle, G. Baxter, & P. Moreno-Ger (Eds.), *Psychology, Pedagogy, and Assessment in Serious Games* (pp. 357–393). Hershey, PA: IGI Global. doi:10.4018/978-1-4666-4773-2.ch017

McQueen, B., Nowak, A., & Olsen, J. K. (2017). Learning by Doing in 3D Environments: Collaborative Efforts in Second Life and Open Sim. In G. Panconesi & M. Guida (Eds.), *Handbook of Research on Collaborative Teaching Practice in Virtual Learning Environments* (pp. 317–347). Hershey, PA: IGI Global. doi:10.4018/978-1-5225-2426-7.ch017

Mena, R. J. (2014). The Quest for a Massively Multiplayer Online Game that Teaches Physics. In T. Connolly, T. Hainey, E. Boyle, G. Baxter, & P. Moreno-Ger (Eds.), *Psychology, Pedagogy, and Assessment in Serious Games* (pp. 292–316). Hershey, PA: IGI Global. doi:10.4018/978-1-4666-4773-2.ch014

Mhiripiri, N. A., & Ureke, O. (2016). Mobile Journalism, Cellphilms, and the Use of the StoryMaker Multimedia Software at a Zimbabwean Media Training University. In L. Briz-Ponce, J. Juanes-Méndez, & F. García-Peñalvo (Eds.), *Handbook of Research on Mobile Devices and Applications in Higher Education Settings* (pp. 318–343). Hershey, PA: IGI Global. doi:10.4018/978-1-5225-0256-2.ch014

Moctezuma, M. S., Ochoa, M. A., & Ramírez, M. S. (2016). Net Generation Features that Enhance Mobile Learning. In L. Briz-Ponce, J. Juanes-Méndez, & F. García-Peñalvo (Eds.), *Handbook of Research on Mobile Devices and Applications in Higher Education Settings* (pp. 92–117). Hershey, PA: IGI Global. doi:10.4018/978-1-5225-0256-2.ch005

Monteiro, F. C., Ribeiro, J., & Martins, R. (2016). Interactive/Automated Method to Count Bacterial Colonies. In J. Rodrigues, P. Cardoso, J. Monteiro, & M. Figueiredo (Eds.), *Handbook of Research on Human-Computer Interfaces, Developments, and Applications* (pp. 556–577). Hershey, PA: IGI Global. doi:10.4018/978-1-5225-0435-1.ch023

Morris, L. R., Morse, C., & Jones, T. K. (2017). How Virtual Work Informs Virtual Learning. In T. Kidd & L. Morris Jr., (Eds.), *Handbook of Research on Instructional Systems and Educational Technology* (pp. 216–227). Hershey, PA: IGI Global. doi:10.4018/978-1-5225-2399-4.ch019

Moseley, A. (2014). A Case for Integration: Assessment and Games. In T. Connolly, T. Hainey, E. Boyle, G. Baxter, & P. Moreno-Ger (Eds.), *Psychology, Pedagogy, and Assessment in Serious Games* (pp. 342–356). Hershey, PA: IGI Global. doi:10.4018/978-1-4666-4773-2.ch016

Mugayitoglu, B., & Kush, J. C. (2017). The Role of Computational Thinking in the Preparation of Pre-Service Teachers. In T. Kidd & L. Morris Jr., (Eds.), *Handbook of Research on Instructional Systems and Educational Technology* (pp. 340–355). Hershey, PA: IGI Global. doi:10.4018/978-1-5225-2399-4.ch029

Mutlu-Bayraktar, D., & Yilmaz, O. (2017). Impact of Kinect Exergame on Mental Computation Speed and Achievement. In T. Kidd & L. Morris Jr., (Eds.), *Handbook of Research on Instructional Systems and Educational Technology* (pp. 186–201). Hershey, PA: IGI Global. doi:10.4018/978-1-5225-2399-4.ch017

Nagata, J. J., Giner, J. R., & Martínez-Abad, F. (2016). Virtual Territorial Heritage in Education through mLearning Resources: Cities of Salamanca Spain and Santiago of Chile. In L. Briz-Ponce, J. Juanes-Méndez, & F. García-Peñalvo (Eds.), *Handbook of Research on Mobile Devices and Applications in Higher Education Settings* (pp. 162–195). Hershey, PA: IGI Global. doi:10.4018/978-1-5225-0256-2.ch008

Naka, T., & Ishida, T. (2016). Dynamic Motion Analysis of Gesture Interaction. In J. Rodrigues, P. Cardoso, J. Monteiro, & M. Figueiredo (Eds.), *Handbook of Research on Human-Computer Interfaces, Developments, and Applications* (pp. 23–51). Hershey, PA: IGI Global. doi:10.4018/978-1-5225-0435-1.ch002

Nazareth, A., Odean, R., & Pruden, S. M. (2017). The Use of Eye-Tracking in Spatial Thinking Research. In C. Was, F. Sansosti, & B. Morris (Eds.), *Eye-Tracking Technology Applications in Educational Research* (pp. 239–260). Hershey, PA: IGI Global. doi:10.4018/978-1-5225-1005-5.ch012

Neri, L., Noguez, J., Whitney, J. M., & Aguilar-Sanchez, G. (2016). Engaging Students to Learn Physics and Mathematics through Short High Quality M-Learning Resources: Design and Implementation Recommendations. In L. Briz-Ponce, J. Juanes-Méndez, & F. García-Peñalvo (Eds.), *Handbook of Research on Mobile Devices and Applications in Higher Education Settings* (pp. 432–452). Hershey, PA: IGI Global. doi:10.4018/978-1-5225-0256-2.ch018

Niblett, S. L., & Rakes, M. L. (2017). Social Media and Technology May Change the Culture of Rape on College Campuses. In T. Kidd & L. Morris Jr., (Eds.), *Handbook of Research on Instructional Systems and Educational Technology* (pp. 263–273). Hershey, PA: IGI Global. doi:10.4018/978-1-5225-2399-4.ch023

Ninaus, M., Witte, M., Kober, S. E., Friedrich, E. V., Kurzmann, J., Hartsuiker, E., & Wood, G. et al. (2014). Neurofeedback and Serious Games. In T. Connolly, T. Hainey, E. Boyle, G. Baxter, & P. Moreno-Ger (Eds.), *Psychology, Pedagogy, and Assessment in Serious Games* (pp. 82–110). Hershey, PA: IGI Global. doi:10.4018/978-1-4666-4773-2.ch005

Nunes, F. B., Herpich, F., Oliveira, M. A., Hannel, K., Flores, M. L., & Valdeni de Lima, J. (2017). A Perspective on the Application of Mastery Learning Theory in Virtual Worlds. In G. Panconesi & M. Guida (Eds.), *Handbook of Research on Collaborative Teaching Practice in Virtual Learning Environments* (pp. 52–75). Hershey, PA: IGI Global. doi:10.4018/978-1-5225-2426-7.ch003

Nunes, F. B., Herpich, F., Voss, G. B., & Medina, R. D. (2016). A Perspective about the Application of Quality of Context in U-Learning Environments. In F. Neto, R. de Souza, & A. Gomes (Eds.), *Handbook of Research on 3-D Virtual Environments and Hypermedia for Ubiquitous Learning* (pp. 410–432). Hershey, PA: IGI Global. doi:10.4018/978-1-5225-0125-1.ch017

Nuninger, W., & Châtelet, J. (2017). Pedagogical Mini-Games Integrated into Hybrid Course to Improve Understanding of Computer Programming: Skill Building Without the Coding Constraints. In R. Alexandre Peixoto de Queirós & M. Pinto (Eds.), *Gamification-Based E-Learning Strategies for Computer Programming Education* (pp. 152–194). Hershey, PA: IGI Global. doi:10.4018/978-1-5225-1034-5.ch008

Nussli, N., & Oh, K. (2017). Field Research in Second Life: Strategies for Discussion Group Facilitation and Benefits of Participation. In G. Panconesi & M. Guida (Eds.), *Handbook of Research on Collaborative Teaching Practice in Virtual Learning Environments* (pp. 348–373). Hershey, PA: IGI Global. doi:10.4018/978-1-5225-2426-7.ch018

O'Shea, P., & Campbell, C. (2016). The Theory and Process Involved with Educational Augmented Reality Game Design. In D. Russell & J. Laffey (Eds.), *Handbook of Research on Gaming Trends in P-12 Education* (pp. 151–163). Hershey, PA: IGI Global. doi:10.4018/978-1-4666-9629-7.ch007

Obilade, T. T. (2016). HCI and E-Learning: Developing a Framework for Evaluating E-Learning. In J. Rodrigues, P. Cardoso, J. Monteiro, & M. Figueiredo (Eds.), *Handbook of Research on Human-Computer Interfaces, Developments, and Applications* (pp. 340–362). Hershey, PA: IGI Global. doi:10.4018/978-1-5225-0435-1.ch014

Obilade, T. T. (2016). Implications of Similarities in Instructional Design, Learner Interface Design and User Interface Design in Designing a User-Friendly Online Module. In J. Rodrigues, P. Cardoso, J. Monteiro, & M. Figueiredo (Eds.), *Handbook of Research on Human-Computer Interfaces, Developments, and Applications* (pp. 321–339). Hershey, PA: IGI Global. doi:10.4018/978-1-5225-0435-1.ch013

Occhioni, M. (2017). Techland: Math and Science in a Virtual World. In G. Panconesi & M. Guida (Eds.), *Handbook of Research on Collaborative Teaching Practice in Virtual Learning Environments* (pp. 407–426). Hershey, PA: IGI Global. doi:10.4018/978-1-5225-2426-7.ch021

Oksanen, K., Lainema, T., & Hämäläinen, R. (2017). Learning from Social Collaboration: A Paradigm Shift in Evaluating Game-Based Learning. In R. Zheng & M. Gardner (Eds.), *Handbook of Research on Serious Games for Educational Applications* (pp. 41–65). Hershey, PA: IGI Global. doi:10.4018/978-1-5225-0513-6.ch003

Oliveira, T. A., Marranghello, N., Silva, A. C., & Pereira, A. S. (2016). Virtual Laboratories Development Using 3D Environments. In F. Neto, R. de Souza, & A. Gomes (Eds.), *Handbook of Research on 3-D Virtual Environments and Hypermedia for Ubiquitous Learning* (pp. 29–53). Hershey, PA: IGI Global. doi:10.4018/978-1-5225-0125-1.ch002

Orr, K., & McGuinness, C. (2014). What is the "Learning" in Games-Based Learning? In T. Connolly, T. Hainey, E. Boyle, G. Baxter, & P. Moreno-Ger (Eds.), Psychology, Pedagogy, and Assessment in Serious Games (pp. 221-242). Hershey, PA: IGI Global. doi:10.4018/978-1-4666-4773-2.ch011

Pacchiega, C. (2017). An Informal Methodology for Teaching Through Virtual Worlds: Using Internet Tools and Virtual Worlds in a Coordinated Pattern to Teach Various Subjects. In G. Panconesi & M. Guida (Eds.), *Handbook of Research on Collaborative Teaching Practice in Virtual Learning Environments* (pp. 163–180). Hershey, PA: IGI Global. doi:10.4018/978-1-5225-2426-7.ch009

Paiva, J. C., Leal, J. P., & Peixoto de Queirós, R. A. (2017). Design and Implementation of an IDE for Learning Programming Languages Using a Gamification Service. In R. Alexandre Peixoto de Queirós & M. Pinto (Eds.), *Gamification-Based E-Learning Strategies for Computer Programming Education* (pp. 295–308). Hershey, PA: IGI Global. doi:10.4018/978-1-5225-1034-5.ch014

Pal, S. (2016). Information Sharing and Communications with Mobile Cloud Technology: Applications and Challenges. In L. Briz-Ponce, J. Juanes-Méndez, & F. García-Peñalvo (Eds.), *Handbook of Research on Mobile Devices and Applications in Higher Education Settings* (pp. 53–71). Hershey, PA: IGI Global. doi:10.4018/978-1-5225-0256-2.ch003

Palilonis, J. A. (2017). Multimedia Active Reading: A Framework for Understanding Learning With Tablet Textbooks. In T. Kidd & L. Morris Jr., (Eds.), *Handbook of Research on Instructional Systems and Educational Technology* (pp. 153–163). Hershey, PA: IGI Global. doi:10.4018/978-1-5225-2399-4.ch014

Panzavolta, S. (2017). Virtual Reality as a Tool for Enhancing Learning in At-Risk Students and Increasing School Inclusion. In G. Panconesi & M. Guida (Eds.), *Handbook of Research on Collaborative Teaching Practice in Virtual Learning Environments* (pp. 242–257). Hershey, PA: IGI Global. doi:10.4018/978-1-5225-2426-7.ch013

Parikh, C. (2017). Eye-Tracking Technology: A Closer Look at Eye-Tracking Paradigms with High-Risk Populations. In C. Was, F. Sansosti, & B. Morris (Eds.), *Eye-Tracking Technology Applications in Educational Research* (pp. 283–302). Hershey, PA: IGI Global. doi:10.4018/978-1-5225-1005-5.ch014

Parlakkılıç, A. (2017). Modular E-Learning Course Design. In T. Kidd & L. Morris Jr., (Eds.), *Handbook of Research on Instructional Systems and Educational Technology* (pp. 228–235). Hershey, PA: IGI Global. doi:10.4018/978-1-5225-2399-4.ch020

Patel, R. P., Lin, J., & Khaderi, S. K. (2017). Beyond Gaming: The Utility of Video Games for Sports Performance. In B. Dubbels (Ed.), *Transforming Gaming and Computer Simulation Technologies across Industries* (pp. 183–191). Hershey, PA: IGI Global. doi:10.4018/978-1-5225-1817-4.ch010

Patronis, M. (2016). The Effect of Using Mobile Devices on Students' Performance in Writing. In L. Briz-Ponce, J. Juanes-Méndez, & F. García-Peñalvo (Eds.), *Handbook of Research on Mobile Devices and Applications in Higher Education Settings* (pp. 250–270). Hershey, PA: IGI Global. doi:10.4018/978-1-5225-0256-2.ch011

Peixoto de Queirós, R. A. (2017). A Survey on Game Backend Services. In R. Alexandre Peixoto de Queirós & M. Pinto (Eds.), *Gamification-Based E-Learning Strategies for Computer Programming Education* (pp. 1–13). Hershey, PA: IGI Global. doi:10.4018/978-1-5225-1034-5.ch001

Pereira, A. S., Moreira, A. A., Chaló, P., Sancho, L., Varela, A., & Oliveira, C. (2016). Development Challenges of a Full Integrated App in Higher Education. In L. Briz-Ponce, J. Juanes-Méndez, & F. García-Peñalvo (Eds.), *Handbook of Research on Mobile Devices and Applications in Higher Education Settings* (pp. 1–24). Hershey, PA: IGI Global. doi:10.4018/978-1-5225-0256-2.ch001

Pereira, G., Brisson, A., Dias, J., Carvalho, A., Dimas, J., Mascarenhas, S., & Paiva, A. et al. (2014). Non-Player Characters and Artificial Intelligence. In T. Connolly, T. Hainey, E. Boyle, G. Baxter, & P. Moreno-Ger (Eds.), *Psychology, Pedagogy, and Assessment in Serious Games* (pp. 127–152). Hershey, PA: IGI Global. doi:10.4018/978-1-4666-4773-2.ch007

Pian, A. (2017). Gamification: Teaching While Playing – Open Issues and Educational Guidelines. In G. Panconesi & M. Guida (Eds.), *Handbook of Research on Collaborative Teaching Practice in Virtual Learning Environments* (pp. 111–128). Hershey, PA: IGI Global. doi:10.4018/978-1-5225-2426-7.ch006

Piecka, D. C., & Calinger, M. (2016). Using a Live Simulation to Teach Human Anatomy and the Diagnostic Process to High School Students. In D. Russell & J. Laffey (Eds.), *Handbook of Research on Gaming Trends in P-12 Education* (pp. 307–325). Hershey, PA: IGI Global. doi:10.4018/978-1-4666-9629-7.ch015

Poitras, E. G., Harley, J. M., Compeau, T., Kee, K., & Lajoie, S. P. (2017). Augmented Reality in Informal Learning Settings: Leveraging Technology for the Love of History. In R. Zheng & M. Gardner (Eds.), *Handbook of Research on Serious Games for Educational Applications* (pp. 272–293). Hershey, PA: IGI Global. doi:10.4018/978-1-5225-0513-6.ch013

Portalés, C., Gimeno, J., Casas, S., Olanda, R., & Giner Martínez, F. (2016). Interacting with Augmented Reality Mirrors. In J. Rodrigues, P. Cardoso, J. Monteiro, & M. Figueiredo (Eds.), *Handbook of Research on Human-Computer Interfaces, Developments, and Applications* (pp. 216–244). Hershey, PA: IGI Global. doi:10.4018/978-1-5225-0435-1.ch009

Proctor, D., & Justice, L. J. (2016). A Future Focus of Gaming: Soft Skills. In D. Russell & J. Laffey (Eds.), *Handbook of Research on Gaming Trends in P-12 Education* (pp. 566–585). Hershey, PA: IGI Global. doi:10.4018/978-1-4666-9629-7.ch027

Raffaeli, L., Montanini, L., Gambi, E., & Spinsante, S. (2016). User Interfaces in Smart Assistive Environments: Requirements, Devices, Applications. In J. Rodrigues, P. Cardoso, J. Monteiro, & M. Figueiredo (Eds.), *Handbook of Research on Human-Computer Interfaces, Developments, and Applications* (pp. 420–443). Hershey, PA: IGI Global. doi:10.4018/978-1-5225-0435-1.ch017

Rahim, K. N., Elamvazuthi, I., Vasant, P., & Ganesan, T. (2016). Robotic Assistive System: Development of a Model based on Artificial Intelligent Technique. In J. Rodrigues, P. Cardoso, J. Monteiro, & M. Figueiredo (Eds.), *Handbook of Research on Human-Computer Interfaces, Developments, and Applications* (pp. 444–477). Hershey, PA: IGI Global. doi:10.4018/978-1-5225-0435-1.ch018

Ramos, C. M., Henriques, C., & Lanquar, R. (2016). Augmented Reality for Smart Tourism in Religious Heritage Itineraries: Tourism Experiences in the Technological Age. In J. Rodrigues, P. Cardoso, J. Monteiro, & M. Figueiredo (Eds.), *Handbook of Research on Human-Computer Interfaces, Developments, and Applications* (pp. 245–272). Hershey, PA: IGI Global. doi:10.4018/978-1-5225-0435-1.ch010

Ratliff, J. A. (2015). A Classification of Video Games and Players. In *Integrating Video Game Research and Practice in Library and Information Science* (pp. 1–21). Hershey, PA: IGI Global. doi:10.4018/978-1-4666-8175-0.ch001

Ratliff, J. A. (2015). Players as People. In *Integrating Video Game Research and Practice in Library and Information Science* (pp. 22–38). Hershey, PA: IGI Global. doi:10.4018/978-1-4666-8175-0.ch002

Remmele, B., & Whitton, N. (2014). Disrupting the Magic Circle: The Impact of Negative Social Gaming Behaviours. In T. Connolly, T. Hainey, E. Boyle, G. Baxter, & P. Moreno-Ger (Eds.), *Psychology, Pedagogy, and Assessment in Serious Games* (pp. 111–126). Hershey, PA: IGI Global. doi:10.4018/978-1-4666-4773-2.ch006

Rezgui, A., Gómez, J. M., & Ben Maaouia, R. (2017). KPI-Based Decision Evaluation System to Enhance QMSs for Higher Educational Institutes. *International Journal of Decision Support System Technology*, *9*(2), 39–55. doi:10.4018/IJDSST.2017040103

Riel, J., Lawless, K. A., & Brown, S. W. (2017). Defining and Designing Responsive Online Professional Development (ROPD): A Framework to Support Curriculum Implementation. In T. Kidd & L. Morris Jr., (Eds.), *Handbook of Research on Instructional Systems and Educational Technology* (pp. 104–115). Hershey, PA: IGI Global. doi:10.4018/978-1-5225-2399-4.ch010

Rique, T. P., Nascimento, S. M., Fujioka, R. D., & da Fonseca de Souza, F. (2016). Architectures for 3D Virtual Environments. In F. Neto, R. de Souza, & A. Gomes (Eds.), *Handbook of Research on 3-D Virtual Environments and Hypermedia for Ubiquitous Learning* (pp. 115–147). Hershey, PA: IGI Global. doi:10.4018/978-1-5225-0125-1.ch005

Roberts, J., & Kidd, T. T. (2017). Millennials, Digital Natives, and the Emergence of New Educational Spaces. In T. Kidd & L. Morris Jr., (Eds.), *Handbook of Research on Instructional Systems and Educational Technology* (pp. 1–10). Hershey, PA: IGI Global. doi:10.4018/978-1-5225-2399-4.ch001

Rodrigues, J., Alves, R., Sousa, L., Negrier, A., Monteiro, J., Cardoso, P., & Bica, P. et al. (2016). PRHOLO: 360° Interactive Public Relations. In J. Rodrigues, P. Cardoso, J. Monteiro, & M. Figueiredo (Eds.), *Handbook of Research on Human-Computer Interfaces, Developments, and Applications* (pp. 162–187). Hershey, PA: IGI Global. doi:10.4018/978-1-5225-0435-1.ch007

Rodrigues, P., & Rosa, P. J. (2017). Eye-Tracking as a Research Methodology in Educational Context: A Spanning Framework. In C. Was, F. Sansosti, & B. Morris (Eds.), *Eye-Tracking Technology Applications in Educational Research* (pp. 1–26). Hershey, PA: IGI Global. doi:10.4018/978-1-5225-1005-5.ch001

Rodríguez, M. A., Juárez, M. Á., Pernas, F. J., Zarzuela, M. M., & Ortega, D. G. (2017). Moodle Game-Based Tool Trivioodle to Support the Learning of Programming Languages and Paradigms. In R. Alexandre Peixoto de Queirós & M. Pinto (Eds.), *Gamification-Based E-Learning Strategies for Computer Programming Education* (pp. 238–260). Hershey, PA: IGI Global. doi:10.4018/978-1-5225-1034-5.ch011

Rosenheck, L., Gordon-Messer, S., Clarke-Midura, J., & Klopfer, E. (2016). Design and Implementation of an MMO: Approaches to Support Inquiry Learning with Games. In D. Russell & J. Laffey (Eds.), *Handbook of Research on Gaming Trends in P-12 Education* (pp. 33–54). Hershey, PA: IGI Global. doi:10.4018/978-1-4666-9629-7.ch002

Rowe, E., Bardar, E., Asbell-Clarke, J., Shane-Simpson, C., & Roberts, S. (2016). Building Bridges: Teachers Leveraging Game-Based Implicit Science Learning in Physics Classrooms. In D. Russell & J. Laffey (Eds.), *Handbook of Research on Gaming Trends in P-12 Education* (pp. 442–468). Hershey, PA: IGI Global. doi:10.4018/978-1-4666-9629-7.ch021

Rudnianski, M., & Kravcik, M. (2014). The Road to Critical Thinking and Intelligence Analysis. In T. Connolly, T. Hainey, E. Boyle, G. Baxter, & P. Moreno-Ger (Eds.), *Psychology, Pedagogy, and Assessment in Serious Games* (pp. 47–61). Hershey, PA: IGI Global. doi:10.4018/978-1-4666-4773-2.ch003

Russell, D. (2016). Flipping the Paradigm of Education: Developing a Comprehensive Educational Program Integrating Virtual Immersive Learning Environments. In D. Russell & J. Laffey (Eds.), *Handbook of Research on Gaming Trends in P-12 Education* (pp. 253–293). Hershey, PA: IGI Global. doi:10.4018/978-1-4666-9629-7.ch013

Russell, D. (2016). The Design of Immersive Virtual Learning Environments Utilizing Problem-Based Learning Templates. In D. Russell & J. Laffey (Eds.), *Handbook of Research on Gaming Trends in P-12 Education* (pp. 105–123). Hershey, PA: IGI Global. doi:10.4018/978-1-4666-9629-7.ch005

Russell, D., & Wallis, S. E. (2016). Designing a Learning Analytic System for Assessing Immersive Virtual Learning Environments. In D. Russell & J. Laffey (Eds.), *Handbook of Research on Gaming Trends in P-12 Education* (pp. 502–526). Hershey, PA: IGI Global. doi:10.4018/978-1-4666-9629-7.ch024

Saberi, M. (2016). Personality-Based Cognitive Design of Characters in Virtual Environments. In J. Turner, M. Nixon, U. Bernardet, & S. DiPaola (Eds.), *Integrating Cognitive Architectures into Virtual Character Design* (pp. 124–150). Hershey, PA: IGI Global. doi:10.4018/978-1-5225-0454-2.ch005

Saleeb, N., Dafoulas, G. A., & Loomes, M. (2016). Personalisation of 3D Virtual Spaces for Enhanced Ubiquitous Learning. In F. Neto, R. de Souza, & A. Gomes (Eds.), *Handbook of Research on 3-D Virtual Environments and Hypermedia for Ubiquitous Learning* (pp. 87–114). Hershey, PA: IGI Global. doi:10.4018/978-1-5225-0125-1.ch004

Sánchez-Prieto, J. C., Olmos-Migueláñez, S., & García-Peñalvo, F. J. (2016). Do Mobile Technologies Have a Place in Universities?: The TAM Model in Higher Education. In L. Briz-Ponce, J. Juanes-Méndez, & F. García-Peñalvo (Eds.), *Handbook of Research on Mobile Devices and Applications in Higher Education Settings* (pp. 25–52). Hershey, PA: IGI Global. doi:10.4018/978-1-5225-0256-2.ch002

Santos, A. (2017). Instructional Strategies for Game-Based Learning. In T. Kidd & L. Morris Jr., (Eds.), *Handbook of Research on Instructional Systems and Educational Technology* (pp. 164–173). Hershey, PA: IGI Global. doi:10.4018/978-1-5225-2399-4.ch015

Santos, B. S., Cardoso, J., Ferreira, B. Q., Ferreira, C., & Dias, P. (2016). Developing 3D Freehand Gesture-Based Interaction Methods for Virtual Walkthroughs: Using an Iterative Approach. In J. Rodrigues, P. Cardoso, J. Monteiro, & M. Figueiredo (Eds.), *Handbook of Research on Human-Computer Interfaces, Developments, and Applications* (pp. 52–72). Hershey, PA: IGI Global. doi:10.4018/978-1-5225-0435-1.ch003

Scheiter, K., & Eitel, A. (2017). The Use of Eye Tracking as a Research and Instructional Tool in Multimedia Learning. In C. Was, F. Sansosti, & B. Morris (Eds.), *Eye-Tracking Technology Applications in Educational Research* (pp. 143–164). Hershey, PA: IGI Global. doi:10.4018/978-1-5225-1005-5.ch008

Shayan, S., Abrahamson, D., Bakker, A., Duijzer, C. A., & van der Schaaf, M. (2017). Eye-Tracking the Emergence of Attentional Anchors in a Mathematics Learning Tablet Activity. In C. Was, F. Sansosti, & B. Morris (Eds.), *Eye-Tracking Technology Applications in Educational Research* (pp. 166–194). Hershey, PA: IGI Global. doi:10.4018/978-1-5225-1005-5.ch009

Simões, A. (2017). Using Game Frameworks to Teach Computer Programming. In R. Alexandre Peixoto de Queirós & M. Pinto (Eds.), *Gamification-Based E-Learning Strategies for Computer Programming Education* (pp. 221–236). Hershey, PA: IGI Global. doi:10.4018/978-1-5225-1034-5.ch010

Singh, A., & Kalyuga, S. (2016). Enhancing the Effectiveness of Educational Hypermedia: A Cognitive Load Approach. In F. Neto, R. de Souza, & A. Gomes (Eds.), *Handbook of Research on 3-D Virtual Environments and Hypermedia for Ubiquitous Learning* (pp. 387–409). Hershey, PA: IGI Global. doi:10.4018/978-1-5225-0125-1.ch016

Singh, L., Thomas, T. D., Gaffar, K. O., & Renville, D. (2016). Mobile Learning among Students and Lecturers in the Developing World: Perceptions Using the UTAUT Model. In L. Briz-Ponce, J. Juanes-Méndez, & F. García-Peñalvo (Eds.), *Handbook of Research on Mobile Devices and Applications in Higher Education Settings* (pp. 402–431). Hershey, PA: IGI Global. doi:10.4018/978-1-5225-0256-2.ch017

Slota, S. T., & Young, M. F. (2017). Stories, Games, and Learning through Play: The Affordances of Game Narrative for Education. In R. Zheng & M. Gardner (Eds.), *Handbook of Research on Serious Games for Educational Applications* (pp. 294–319). Hershey, PA: IGI Global. doi:10.4018/978-1-5225-0513-6.ch014

Smart, P. R., Scutt, T., Sycara, K., & Shadbolt, N. R. (2016). Integrating ACT-R Cognitive Models with the Unity Game Engine. In J. Turner, M. Nixon, U. Bernardet, & S. DiPaola (Eds.), *Integrating Cognitive Architectures into Virtual Character Design* (pp. 35–64). Hershey, PA: IGI Global. doi:10.4018/978-1-5225-0454-2.ch002

Smith, P. A., & Welch, S. (2016). The Use of Smart Toys in Learning Games. In D. Russell & J. Laffey (Eds.), *Handbook of Research on Gaming Trends in P-12 Education* (pp. 470–480). Hershey, PA: IGI Global. doi:10.4018/978-1-4666-9629-7.ch022

Snyder, A. E., & Van Voorhis, V. (2016). Playful Transition: Serious Games for the Middle School Years. In D. Russell & J. Laffey (Eds.), *Handbook of Research on Gaming Trends in P-12 Education* (pp. 195–221). Hershey, PA: IGI Global. doi:10.4018/978-1-4666-9629-7.ch010

Sousa, L., Figueiredo, M., Monteiro, J., Bidarra, J., Rodrigues, J., & Cardoso, P. (2016). Developments of Serious Games in Education. In J. Rodrigues, P. Cardoso, J. Monteiro, & M. Figueiredo (Eds.), *Handbook of Research on Human-Computer Interfaces, Developments, and Applications* (pp. 392–419). Hershey, PA: IGI Global. doi:10.4018/978-1-5225-0435-1.ch016

Stover, S., Patitu, C., & DuVivier, R. (2016). Student Use of Cell Phones to Conduct Real-Time Polling. In L. Briz-Ponce, J. Juanes-Méndez, & F. García-Peñalvo (Eds.), *Handbook of Research on Mobile Devices and Applications in Higher Education Settings* (pp. 271–296). Hershey, PA: IGI Global. doi:10.4018/978-1-5225-0256-2.ch012

Suovuo, T. "., Lahti, I., & Smed, J. (2016). Game Design Frameworks and Reality Guides. In D. Russell, & J. Laffey (Eds.), Handbook of Research on Gaming Trends in P-12 Education (pp. 85-104). Hershey, PA: IGI Global. doi:10.4018/978-1-4666-9629-7.ch004

Swacha, J. (2017). Scripting Environments of Gamified Learning Management Systems for Programming Education. In R. Alexandre Peixoto de Queirós & M. Pinto (Eds.), *Gamification-Based E-Learning Strategies for Computer Programming Education* (pp. 278–294). Hershey, PA: IGI Global. doi:10.4018/978-1-5225-1034-5.ch013

Talbot, T. B. (2017). Making Lifelike Medical Games in the Age of Virtual Reality: An Update on "Playing Games with Biology" from 2013. In B. Dubbels (Ed.), *Transforming Gaming and Computer Simulation Technologies across Industries* (pp. 103–119). Hershey, PA: IGI Global. doi:10.4018/978-1-5225-1817-4.ch006

Taub, M., Martin, S. A., Azevedo, R., & Mudrick, N. V. (2016). The Role of Pedagogical Agents on Learning: Issues and Trends. In F. Neto, R. de Souza, & A. Gomes (Eds.), *Handbook of Research on 3-D Virtual Environments and Hypermedia for Ubiquitous Learning* (pp. 362–386). Hershey, PA: IGI Global. doi:10.4018/978-1-5225-0125-1.ch015

Tavernise, A., & Bertacchini, F. (2016). Designing Educational Paths in Virtual Worlds for a Successful Hands-On Learning: Cultural Scenarios in NetConnect Project. In F. Neto, R. de Souza, & A. Gomes (Eds.), *Handbook of Research on 3-D Virtual Environments and Hypermedia for Ubiquitous Learning* (pp. 148–167). Hershey, PA: IGI Global. doi:10.4018/978-1-5225-0125-1.ch006

Tegon, R. (2017). Design Learning for Inclusion in Virtual World Through Cognitive Enhancement and the Feuerstein MLE's Perspective. In G. Panconesi & M. Guida (Eds.), *Handbook of Research on Collaborative Teaching Practice in Virtual Learning Environments* (pp. 181–199). Hershey, PA: IGI Global. doi:10.4018/978-1-5225-2426-7.ch010

Thompson, K., & Markauskaite, L. (2014). Identifying Group Processes and Affect in Learners: A Holistic Approach to Assessment in Virtual Worlds in Higher Education. In S. Kennedy-Clark, K. Everett, & P. Wheeler (Eds.), *Cases on the Assessment of Scenario and Game-Based Virtual Worlds in Higher Education* (pp. 175–210). Hershey, PA: IGI Global. doi:10.4018/978-1-4666-4470-0.ch006

Thompson, N., Peppler, K., & Danish, J. (2017). Designing BioSim: Playfully Encouraging Systems Thinking in Young Children. In R. Zheng & M. Gardner (Eds.), *Handbook of Research on Serious Games for Educational Applications* (pp. 149–167). Hershey, PA: IGI Global. doi:10.4018/978-1-5225-0513-6.ch007

Tran, B. (2014). Rhetoric of Play: Utilizing the Gamer Factor in Selecting and Training Employees. In T. Connolly, T. Hainey, E. Boyle, G. Baxter, & P. Moreno-Ger (Eds.), *Psychology, Pedagogy, and Assessment in Serious Games* (pp. 175–203). Hershey, PA: IGI Global. doi:10.4018/978-1-4666-4773-2.ch009

Tricarico, M. (2017). Virtual Worlds: Places Where You Can Find Your True Talents. In G. Panconesi & M. Guida (Eds.), *Handbook of Research on Collaborative Teaching Practice in Virtual Learning Environments* (pp. 549–568). Hershey, PA: IGI Global. doi:10.4018/978-1-5225-2426-7.ch027

Trinchero, R. (2017). Learn from Experience to Build Competence: A Model for Structuring Active Learning Practices in Virtual Learning Environments. In G. Panconesi & M. Guida (Eds.), *Handbook of Research on Collaborative Teaching Practice in Virtual Learning Environments* (pp. 1–14). Hershey, PA: IGI Global. doi:10.4018/978-1-5225-2426-7.ch001

Turkay, S., & Kinzer, C. K. (2017). The Relationship between Avatar-Based Customization, Player Identification, and Motivation. In B. Dubbels (Ed.), *Transforming Gaming and Computer Simulation Technologies across Industries* (pp. 48–79). Hershey, PA: IGI Global. doi:10.4018/978-1-5225-1817-4.ch003

Turner, J. O. (2016). Virtual Soar-Agent Implementations: Examples, Issues, and Speculations. In J. Turner, M. Nixon, U. Bernardet, & S. DiPaola (Eds.), *Integrating Cognitive Architectures into Virtual Character Design* (pp. 181–212). Hershey, PA: IGI Global. doi:10.4018/978-1-5225-0454-2.ch007

Usart, M., & Romero, M. (2014). Time Factor Assessment in Game-Based Learning: Time Perspective and Time-on-Task as Individual Differences between Players. In T. Connolly, T. Hainey, E. Boyle, G. Baxter, & P. Moreno-Ger (Eds.), *Psychology, Pedagogy, and Assessment in Serious Games* (pp. 62–81). Hershey, PA: IGI Global. doi:10.4018/978-1-4666-4773-2.ch004

Ustun, V., & Rosenbloom, P. S. (2016). Towards Truly Autonomous Synthetic Characters with the Sigma Cognitive Architecture. In J. Turner, M. Nixon, U. Bernardet, & S. DiPaola (Eds.), *Integrating Cognitive Architectures into Virtual Character Design* (pp. 213–237). Hershey, PA: IGI Global. doi:10.4018/978-1-5225-0454-2.ch008

Vahldick, A., Marcelino, M. J., & Mendes, A. J. (2017). Principles of a Casual Serious Game to Support Introductory Programming Learning in Higher Education. In R. Alexandre Peixoto de Queirós & M. Pinto (Eds.), *Gamification-Based E-Learning Strategies for Computer Programming Education* (pp. 53–78). Hershey, PA: IGI Global. doi:10.4018/978-1-5225-1034-5.ch004

Van Eaton, G., & Clark, D. B. (2016). Designing Digital Objects to Scaffold Learning. In D. Russell & J. Laffey (Eds.), *Handbook of Research on Gaming Trends in P-12 Education* (pp. 237–252). Hershey, PA: IGI Global. doi:10.4018/978-1-4666-9629-7.ch012

van Rosmalen, P., Wilson, A., & Hummel, H. G. (2014). Games for and by Teachers and Learners. In T. Connolly, T. Hainey, E. Boyle, G. Baxter, & P. Moreno-Ger (Eds.), *Psychology, Pedagogy, and Assessment in Serious Games* (pp. 243–269). Hershey, PA: IGI Global. doi:10.4018/978-1-4666-4773-2.ch012

von Gillern, S. (2016). Games and Their Embodied Learning Principles in the Classroom: Connecting Learning Theory to Practice. In D. Russell & J. Laffey (Eds.), *Handbook of Research on Gaming Trends in P-12 Education* (pp. 55–84). Hershey, PA: IGI Global. doi:10.4018/978-1-4666-9629-7.ch003

Vu, P., Fredrickson, S., Hoehner, P., & Ziebarth-Bovill, J. K. (2016). An Examination of Pre-Service Teachers' Attitudes towards Game-Based Learning. In D. Russell & J. Laffey (Eds.), *Handbook of Research on Gaming Trends in P-12 Education* (pp. 294–305). Hershey, PA: IGI Global. doi:10.4018/978-1-4666-9629-7.ch014

Vu, P., & Gaskill, M. (2016). Can Pre-Service Teachers Create Digital Game-Based Activities without Coding Knowledge? In D. Russell & J. Laffey (Eds.), *Handbook of Research on Gaming Trends in P-12 Education* (pp. 380–393). Hershey, PA: IGI Global. doi:10.4018/978-1-4666-9629-7.ch018

Wang, Y. F. (2017). Teacher Training in 3D Virtual Worlds: Understanding Immersive Learning for Teaching Practices. In G. Panconesi & M. Guida (Eds.), *Handbook of Research on Collaborative Teaching Practice in Virtual Learning Environments* (pp. 200–220). Hershey, PA: IGI Global. doi:10.4018/978-1-5225-2426-7.ch011

Webster, K. S. (2017). RSS and Syndication for Educators. In T. Kidd & L. Morris Jr., (Eds.), *Handbook of Research on Instructional Systems and Educational Technology* (pp. 366–379). Hershey, PA: IGI Global. doi:10.4018/978-1-5225-2399-4.ch031

Wexler, B. E. (2017). Computer-Presented and Physical Brain-Training Exercises for School Children: Improving Executive Functions and Learning. In B. Dubbels (Ed.), *Transforming Gaming and Computer Simulation Technologies across Industries* (pp. 206–224). Hershey, PA: IGI Global. doi:10.4018/978-1-5225-1817-4.ch012

Yu, C., & Franz, D. P. (2017). Visiting Technological Pedagogical and Content Knowledge (TPACK): Issues and Challenges for Teachers' Professional Development. In T. Kidd & L. Morris Jr., (Eds.), *Handbook of Research on Instructional Systems and Educational Technology* (pp. 380–391). Hershey, PA: IGI Global. doi:10.4018/978-1-5225-2399-4.ch032

Zamora, R., Vélez, J., & Villa, J. L. (2016). Contributions of Collaborative and Immersive Environments in Development a Remote Access Laboratory: From Point of View of Effectiveness in Learning. In F. Neto, R. de Souza, & A. Gomes (Eds.), *Handbook of Research on 3-D Virtual Environments and Hypermedia for Ubiquitous Learning* (pp. 1–28). Hershey, PA: IGI Global. doi:10.4018/978-1-5225-0125-1.ch001

Zanin, M., & Papo, D. (2016). Identity Assurance through EEG Recordings. In J. Rodrigues, P. Cardoso, J. Monteiro, & M. Figueiredo (Eds.), *Handbook of Research on Human-Computer Interfaces, Developments, and Applications* (pp. 545–555). Hershey, PA: IGI Global. doi:10.4018/978-1-5225-0435-1.ch022

Zelinski, E. M. (2017). Games and Other Training Interventions to Improve Cognition in Healthy Older Adults. In B. Dubbels (Ed.), *Transforming Gaming and Computer Simulation Technologies across Industries* (pp. 192–205). Hershey, PA: IGI Global. doi:10.4018/978-1-5225-1817-4.ch011

Zentall, S. R., & Junglen, A. G. (2017). Investigating Mindsets and Motivation through Eye Tracking and Other Physiological Measures. In C. Was, F. Sansosti, & B. Morris (Eds.), *Eye-Tracking Technology Applications in Educational Research* (pp. 48–64). Hershey, PA: IGI Global. doi:10.4018/978-1-5225-1005-5.ch003

Zheng, R. Z., & Truong, T. N. (2017). A Framework for Promoting Knowledge Transfer in SNS Game-Based Learning. In R. Zheng & M. Gardner (Eds.), *Handbook of Research on Serious Games for Educational Applications* (pp. 66–91). Hershey, PA: IGI Global. doi:10.4018/978-1-5225-0513-6.ch004

Zook, A. (2016). Game AGI beyond Characters. In J. Turner, M. Nixon, U. Bernardet, & S. DiPaola (Eds.), *Integrating Cognitive Architectures into Virtual Character Design* (pp. 266–293). Hershey, PA: IGI Global. doi:10.4018/978-1-5225-0454-2.ch010

Zumbach, J., & Moser, S. (2016). Examining the Effectiveness of Hyperaudio Learning Environments. In F. Neto, R. de Souza, & A. Gomes (Eds.), *Handbook of Research on 3-D Virtual Environments and Hypermedia for Ubiquitous Learning* (pp. 433–450). Hershey, PA: IGI Global. doi:10.4018/978-1-5225-0125-1.ch018

About the Author

Pio Alfredo Di Tore, PhD, is teacher of Italian literature in high schools. For several years, he has worked as a software designer for software houses specialized in eLearning solutions. His research interests are focused primarily on the relationship between human movement sciences, technology and cognition, with particular attention to natural interfaces and simulation environments.

Index

www.ingramcontent.com/pod-product-compliance
Ingram Content Group UK Ltd.
Pitfield, Milton Keynes, MK11 3LW, UK
UKHW050944290926
15470UKWH00005B/1632

9 781522 524557